Springer Proceedings in Earth and Environmental Sciences

The series Springer Proceedings in Earth and Environmental Sciences publishes proceedings from scholarly meetings and workshops on all topics related to Environmental and Earth Sciences and related sciences. This series constitutes a comprehensive up-to-date source of reference on a field or subfield of relevance in Earth and Environmental Sciences. In addition to an overall evaluation of the interest, scientific quality, and timeliness of each proposal at the hands of the publisher, individual contributions are all refereed to the high quality standards of leading journals in the field. Thus, this series provides the research community with well-edited, authoritative reports on developments in the most exciting areas of environmental sciences, earth sciences and related fields.

Jan Baeyens · Raf Dewil · Barbara Rossi ·
Yimin Deng
Editors

New Developments in Environmental and Energy Technologies

Proceedings of the 2023 5th International Conference on Environmental Sciences and Renewable Energy

Springer

Editors
Jan Baeyens
Beijing University of Chemical Technology
Beijing, China

Raf Dewil
Department of Chemical Engineering
KU Leuven
Leuven, Belgium

Barbara Rossi
Oxford University
Oxford, UK

Yimin Deng
KU Leuven
Sint-Katelijne-Waver, Belgium

ISSN 2524-342X ISSN 2524-3438 (electronic)
Springer Proceedings in Earth and Environmental Sciences
ISBN 978-981-96-4344-8 ISBN 978-981-96-4345-5 (eBook)
https://doi.org/10.1007/978-981-96-4345-5

This Springer imprint is published by the registered company Springer Nature Singapore Pte Ltd.
The registered company address is: 152 Beach Road, #21-01/04 Gateway East, Singapore 189721, Singapore

Preface

It was our pleasure and honor to host the 2023 International Conference on Environment Sciences and Renewable Energy (ESRE 2023), held in Vienna, Austria, during June 19–21, 2023. The conference has emerged as an exclusive opportunity for the participants from all over the globe to present and discuss their recent research results in the respective environmental sciences and renewable energy areas.

The international conference was divided into two major parts, including keynote speeches and oral presentations. We welcomed over 40 leading environmental and energy researchers, engineers, and scientists from numerous countries, including Austria, Belgium, UK, Portugal, Japan, Germany, South Korea, Thailand, Burkina Faso, India, Czech Republic, Philippines, USA, Qatar, Oman, Libya, France, China, Peru, etc.

Prof. Franz Winter from the Vienna University of Technology, Austria, delivered a keynote speech with the title "The TU Wien CO_2 Refinery—from CO_2 to Products". Prof. Lise Appels from KU Leuven, Belgium, presented her keynote speech with the title "Organic Waste as Key-feedstock in a Sustainable Bio-economy: Smart Resource Recovery for Energy, Materials and Chemicals Supply". Prof. Xianfeng Fan from the University of Edinburgh, UK, delivered keynote speech "Novel Bi-solvent Non-aqueous AMP Absorbents for CO_2 Capture". Finally, Dr. Yimin Deng from the Beijing Institute of Technology, China, presented her keynote review about "Solar-driven Hydrogen Production: Recent Research Results". Many researchers, engineers, academics as well as industrials from all over the world have presented their research results during the session presentations. Topics included Renewable Energy Generation and Transformation, Energy Market and Industrial Computer Vision, Clean Energy Technology and Energy Management, Environmental Pollution Monitoring, Analysis and Control, and Pollution Abatement Technologies. All presentations were oral with ample time given for Questions and Answers at the end of the presentation. One presentation per session was selected as "best presentation" for its generally outstanding quality and scientific contribution.

The present proceedings present a selection of papers submitted to the conference. All papers were subjected to peer review by conference committee members and international reviewers. The papers were selected on the basis of their quality and their relevancy to the conference. The volume tends to present to the readers the recent advances in the field of Environment Sciences and Renewable Energy. The different papers were grouped per related areas and presented in four sections, including renewable energy and energy management, sustainable development, advanced engineering techniques, environmental technologies.

We would like to express our great appreciation to the organizing committee, reviewers, speakers, chairpersons, sponsors, and all the conference participants for their support to ESRE 2023. Without their participation, it would have been impossible to hold ESRE 2023 successfully and to ensure a high quality of papers for the conference proceeding.

Beijing, China

Jan Baeyens
General Chair, ESRE 2023

Committees

Honorary Chair

Wilhelm Höflinger	Vienna University of Technology, Austria

General Chair

Jan Baeyens	Beijing University of Chemical Technology, China; KU Leuven, Belgium

Conference Chairs

Raf Dewil	KU Leuven, Belgium
Lise Appels	KU Leuven, Belgium

Program Chairs

Huili Zhang	Beijing University of Chemical Technology, China
Germán Mazza	University of Neuquèn, Argentina
Xienfeng Fan	University of Edinburgh, U.K.
Franz Winter	Vienna University of Technology, Austria
Martin Schiemann	Bochum University, Germany
Kazuyki Oshida	University of Kyoto, Japan
John Wheeldon	Electric Power Research Institute, Palo Alto, USA
Yaning Zhang	Harbin Institute of Technology, China
Hongwei Wu	University of Hertfordshire, United Kingdom

Technical Committee Members

Catalina Iticescu	University of Galati, Romania
Md. Hasanuzzaman	University of Malaya, Malaysia
Cao Yanpeng	Zhejiang University, China
Hongwei Wu	University of Hertfordshire, United Kingdom

Contents

Environmental Pollution Treatment and Resource Management

Hydraulic Operation and Aeration Variation in COD Removal in Oxidation Ditches in High Andean Climate

Martin Huaman Carranza[1,2](✉), Kiko Depaz Celi[2], María Pasco Meza[2], and Rosario Polo Salazar[2]

[1] Yaku Taita Research and Innovation, AT 401, Huaraz, Peru
mhuamanc@unasam.edu.pe
[2] Santiago Antúnez de Mayolo University, Independencia AVC 200, Huaraz, Peru

Abstract. The objective was to demonstrate the reduction of organic matter in the oxidation ditch with a design flow of 0.519 m3d-1, total volume of 0.368 m3, considering a hydraulic retention time of 17 h, aeration with a brush was produced. Surface swivel, with electric motor of 0.5 horsepower. The results of the hydraulic conditions were associated with temperature, pH, dissolved oxygen, COD and BOD, the relationship of recirculation flow and inlet flow of 0.752 was reported. The temperature reached 18.55 °C in the tributary, in the aeration zone it reached 17.08 °C and varying at the outlet with an amount of 17.9 °C as a minimum value; the pH oscillated between 7.42 to 7.92 in the influent, 7.86 to 8.22 in the reactor and 7.81 to 8.23 in the effluent; dissolved oxygen in the reactor was between 0.5 mg/L and 2.0 mg/L, guaranteeing aerobic oxidation; It was possible to verify that the average removal efficiencies of the biochemical an chemical oxygen demand were 77.76%, and performing the confidence test, a value of 99.99% was determined for inlet and outlet flows. It is concluded that there is a relationship between good hydraulic control and the efficiency of the oxidation ditch as a secondary treatment unit.

Keywords: Oxidation ditch · biochemical oxygen demand · operation · wastewater · hydraulic retention time

1 Introduction

Domestic wastewater created in underdeveloped nations is released directly into bodies of water without proper treatment [1], posing possible risks to the environment and public health of adjacent populations [2], As peri-urban and urban populations increase, there is a need to expand the sanitary sewer networks at reasonable costs [3], This includes upgrading conventional wastewater treatment plants to further reduce pollution in family environments and the risk of catching water-borne illnesses [4]. The activated sludge process, a sophisticated system that involves several bacterial conversion and transport stages, is an alternative for treating wastewater, [5]. It also follows that these facts must be taken into account to lessen the problems in design and operation where

J. Baeyens et al. (Eds.): ESRE 2023, pp. 3–13, 2025.
https://doi.org/10.1007/978-981-96-4345-5_1

there are demands for effluents that require the removal of nitrogen and phosphorus from the wastewater treatment plants (WWTP) and which ensures an absorption of organic matter [6]. Particularly with regard to biological nitrogen removal, overall nitrogen removal is typically unreliable due to the restricted carbon sources for denitrification [7]. In certain situations, the operation and design are complicated by the simultaneous biological removal of phosphorus and nitrogen from the activated sludge [8]. The removal of COD that is suspended together with nitrogen and organic substances is one of the important features of the oxidation ditch process [9]. To describe this phenomenon, synthetic kinetic modeling including BOD oxidation rates, nitrification, denitrification, DO and alkalinity changes, and sludge growth are proposed [10]. To treat domestic wastewater for small to moderate wastewater treatme,nt plants, oxidation ditches (OD) are widely used for their simple construction and operation, and for their treatment efficiency [11].

Combining alternating aerobic and anoxic conditions in the ditch can simultaneously remove BOD and N [12]. Adding an upstream anaerobic zone will enhance the biological P removal [13]. Since the nutrient removal efficiency is hampered by fluctuating influent parameters [14], the measurement of these parameters and their corresponding mass balances under long-duration steady operating conditions will provide a better definition of the system reliability, efficiency, and underlying mechanisms [16, 17]. The mass balances must also account for pollutant accumulation within the OD [18]. An identification of operating conditions in DO associate with the removal of organic load was developed where an anoxic stage and a dissolved oxygen (DO) control technology are used for the biological removal of organic matter [19]. By independently monitoring and controlling the airflow and recirculation flow rate based on DO values at identified points along the trench, it is possible to verify removal is taking place in clear, clear aerobic and anoxic zones stable [20]. The control of the oxidation ditch between the hydraulic operation and the BOD removal efficiency can be compared and analyzed by means of the simulation of the ASM2d model [21]. It is known that there are obvious differences in organic matter removal performance for oxygenation and nitrogen capacity between oxidation ditch technology when flow levels and retention times vary during the day [22]. it is highlighted that in the stable aerobic and anoxic zones in the channel a lower concentration of nitrate and ammonia in the effluent can be achieved, for which reason with an effective operation control of this technology the BOD reduction yields are very good [22].

Evaluation of oxidation ditch technologies based on hydraulics and treatment performance also focuses on organic loading and hydraulic loading mechanisms, which go hand in hand with removal and efficiency of carbon source consumption to through mass balance analysis [23]. The treatment systems, the more associated they are in control and operation, the yields will be optimal [24]. The important thing is to know that there is an effect of the hydraulic load on the removal of contaminants, and associating the organic load and adequate retention times if the removal of contaminants is achieved effectively in the oxidation ditches [25].

2 Methods and Procedure

2.1 Pilot Design

For the design of the cylindrical-based carousel-type Oxidation Ditch [26], and the raw residual water that reaches the Tuyu Ruri experimental center at 2590 m above sea level was characterized.

The reactor has two curved vertical separators, a vertical separator, a sludge recirculation pipe, and a rotating brush surface aerator. To obtain the cylindrical base, the following criteria were considered. As can be seen in Fig. 1 [27].

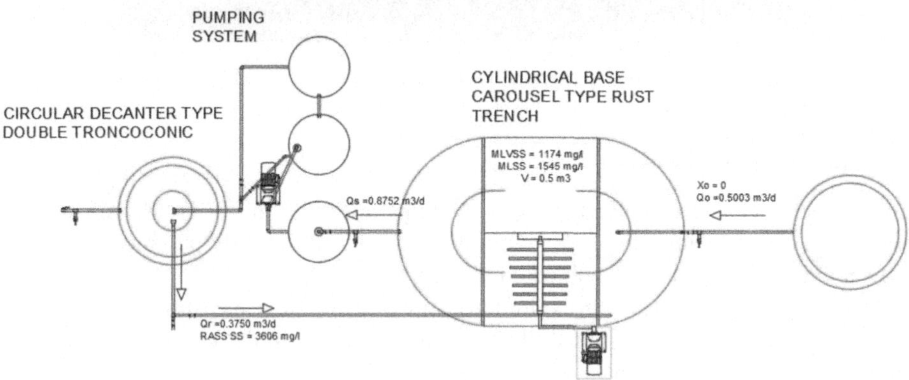

Fig. 1. Oxidation ditch pilot treatment plant

For the rotating brush surface aerator, flat wire was used which were each cut to 21 cm, then bent at one end 2 cm to give it an "L" shape. The rotating surface aerator was installed. It is made up of 10 groups of brushes and each group has 3 blades. To prevent corrosion, sanding was carried out, painted with gray anticorrosive enamel, and bathed with light blue enamel. The pilot installation process can be seen in Fig. 2.

2.2 Monitoring and Control

During the adaptation stage that lasted 1.5 months and a half, monitored the main field parameters: temperature, pH, dissolved oxygen. In the treatment stage, the monitoring of the physicochemical parameters was carried out: temperature, pH, dissolved oxygen, total suspended solids, biochemical oxygen demand, ammonium, nitrite, and nitrate, for a period of four months. 04 monitoring points were considered: Inlet to secondary treatment, Oxidation Ditch aeration zone and Outlet of the double frustoconical circular decanter, for the operation of the carousel-type secondary treatment, used 0.519 m3/d. The type of sampling used is non-probabilistic.

Fig. 2. Installation, testing and commissioning of pilot oxidation ditch

2.3 Operation and Function

Temperature, pH and dissolved oxygen readings are taken in the field with already standardized equipment complying with all protocols [28], while parameters such as BOD, solids and nutrients are subjected to laboratory tests.

In the hydraulic control, the volumetric method for flow measurement was considered, tracers were used to check the retention time and at the end the evidence the relationship with mass balances and tracers.

3 Results and Discussion

The reactor design presents a design flow of 0.519 m^3d^{-1}, a volume of the spherical cap of 0.155 m^3, an ellipse volume of 0.213 m^3, a volume of the Oxidation Ditch of 0.368 m^3 and a hydraulic retention time of 17 h, also has the oxidation ditch has a width of 0.50m, a depth of 0.35m, a length of the vertical separator of 0.80m and a total length of 1.80m. The carousel-type oxidation ditch was used, it has an inlet pipe, an electrical panel, two curved vertical separators, a vertical separator, a sludge recirculation pipe, a rotary brush surface aerator, a 0.5-phase single-phase electric motor. HP and an outlet pipe.

As can be seen in Fig. 3, the different quantities of inlet flow to the oxidation ditch treatment plant were verified, reaching average values of 0.519 m^3d^{-1} for the entire sampling, a maximum value of 0.55 m^3/d and a minimum flow of 0.48 m^3/d. d. achieving regular flows throughout the process and having a production of 3.633 m^3/week (Table 1).

In Table 1, the pH control inside the oxidation box pilot is important to guarantee that biological removal is correctly promoted, giving conditions to aerobic and anoxic microorganisms [27]. it has been verified that the pH level inside the oxidation ditch pilot is 7.6277 and the last month a value of 8.045 was achieved, this variation is common for water from sanitary sewers. It was also deduced that in the oxidation ditch it presented a minimum average pH of 7.42 and a maximum average pH of 8.23, at moments where flow variation was evident during all the monitoring campaigns. Table 2 also considered

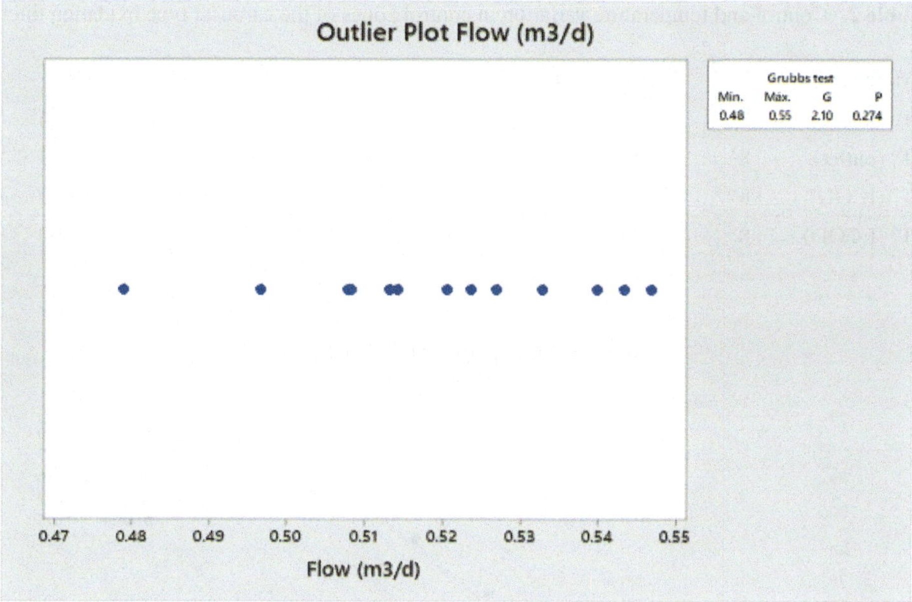

Fig. 3. Control and variation of flow during hours of the process

Table 1. pH control inside the oxidation box pilot

Variable	N	Mean	Std. Dev	Mín	Máx	G	P
pH (Inlet)	13	7.6277	0.1719	7.4200	7.9200	1.70	0.966
pH (ICOD)	13	8.0177	0.1031	7.8700	8.1900	1.67	1.000
pH (ECOD)	13	8.0446	0.1144	7.8600	8.2200	1.61	1.000
pH (outle)	13	8.0423	0.1197	7.8100	8.2300	1.94	0.473

the behavior of the temperature at the inlet flow and outlet flow and was also measured in two oxygenation zones of the carousel ditch, internal compartment (ICOD) and external compartment of the oxidation ditch (ECOD), it was possible to obtain registered values with a variation of 14.7 to 20.2 °C.

The amount of dissolved oxygen present in the waters inside the tank was determined by taking several specific samples to know its variation, being able to quantify it at different times.

Figure 4 shows the variation of dissolved oxygen in the oxidation ditch shows an average value of 1.19 mg/l, having a maximum of 2.0 mg/l in the initial stage of adaptation and reached a minimum value of 0.5 mg/l in per the arrival load within the drainage network, even so, achieving a confidence level of 95%; ensuring that the variation has the correct aeration capacity for the aerobic process.

Table 2. Control and temperature variation in control zones of the carousel type oxidation ditch

Variable	N	Mean	Std. Dev	Mín	Máx	G	P
T° (inlet)	8	18.551	1.608	16.010	20.200	1.58	0.706
T° (outlet)	8	17.362	1.708	14.700	19.300	1.56	0.753
T° (ICOD)	8	17.063	1.684	14.700	19.200	1.40	1.000
T° (ECOD)	8	17.088	1.755	14.800	19.200	1.30	1.000

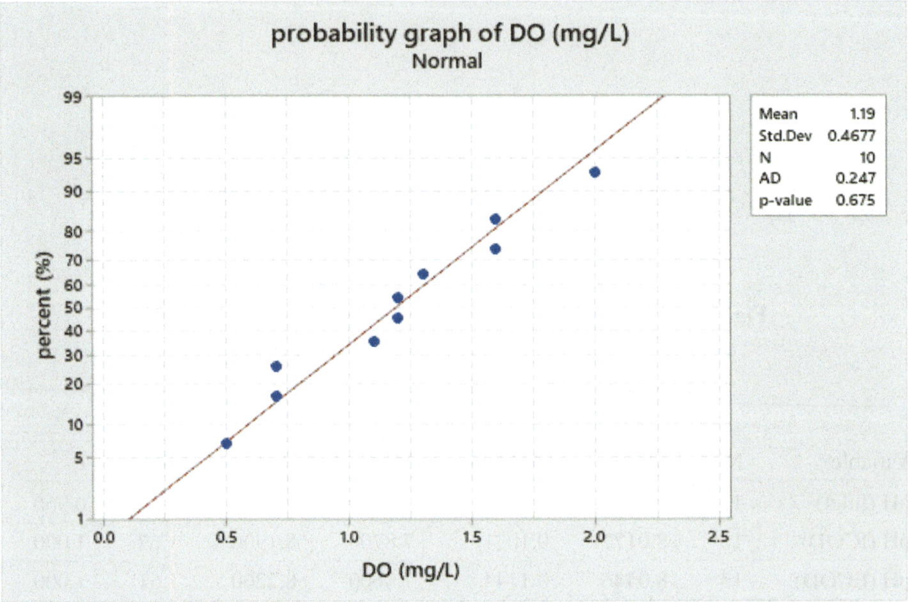

Fig. 4. Report of quantity and fluctuation of dissolved oxygen found in the carousel type tank of the oxidation ditch.

To know the hydraulic behavior, the verification of the recirculation[29] and output flow to the secondary clarifier was carried out, achieving proportionality relationships between said flows to know the real factor of operation in the oxidation ditch as well as shown in Fig. 5 the amounts of flows measured at different operating times. With the average flow data obtained during the entire operation time, the relationship $Qr/Qi = 0.752$, it was also possible to calculate the relationship of $Qr/Qo = 0.433$, this means that if there is a difference in the flow balances in the ditch when the flow returns and when it goes out to the decanter, achieving with these an acceptable performance of operation.

The flows measured in the oxygenation zone of the trench show a constant value that recirculates throughout the operation time, reaching an average of 0.372 m^3/d and a maximum of 0.381 m^3/d, achieving that the output of the typical trench Carousel also generated a continuous average decanter discharge of 0.869 m^3/d, having an adequate control of the hydraulic behavior inside the reactor.

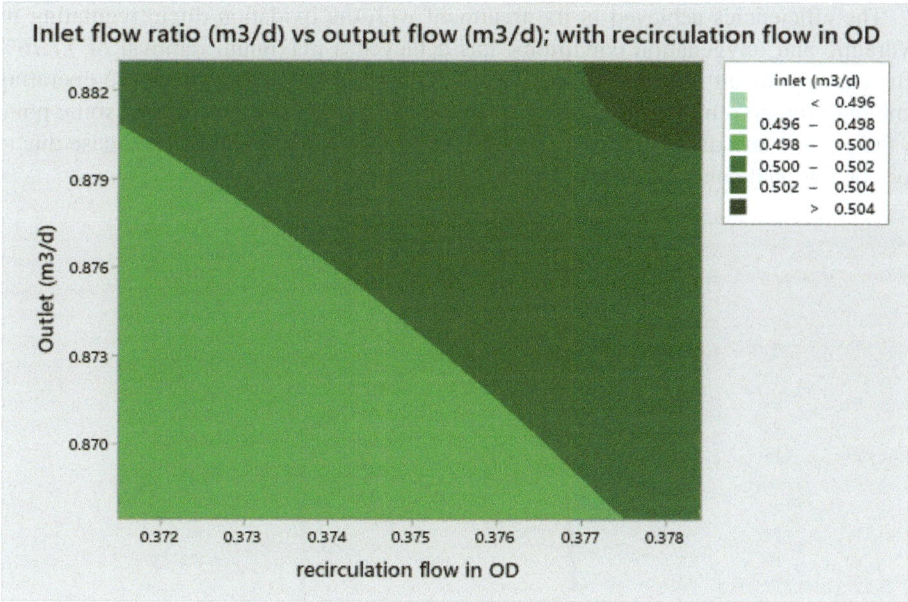

Fig. 5. Amount of input, output, and recirculation operation flows

With these data obtained, we related it to the amount of dissolved oxygen in the two internal and external zones of the ditch, reaching a maximum oxygen load of 1.1 g/day, therefore, the maximum consumption studied reached 7.7 g/week.

Once the flow rates were regulated and knowing the amount of dissolved oxygen that is counted in the oxidation ditch, the amounts of COD removal began to be measured with the already permanent and continuous calibrations, now in Table 3 of samples the removals obtained, and the efficiencies achieved.

Table 3. BOD removal efficiencies by checking aeration and flow rate

Samples	Inlet COD (mg/l)	Outlet COD (mg/l)	Efficiency (%)
M1	285.10	63.40	77.76%
M2	298.80	68.70	77.01%
M3	295.40	69.30	76.54%
M4	125.30	48.00	61.69%
M5	115.90	49.00	57.72%
M6	102.40	50.00	51.17%
M7	78.80	75.00	4.82%
M8	84.30	80.20	4.86%
M9	85.10	82.10	3.53%

The efficiencies achieved in the treatment with the oxidation ditch, regulating its hydraulic and oxygenation conditions, has achieved a maximum removal of 77.76% efficiency, reducing from 285.10 mg/l to 63.40 mg/l of COD, in most of the operating time the average removal efficiency was maintained at 66.98%, existing at some point in the process and disproportion of up to 3.53% but considered an isolated case due to poor operational control in the WWTP.

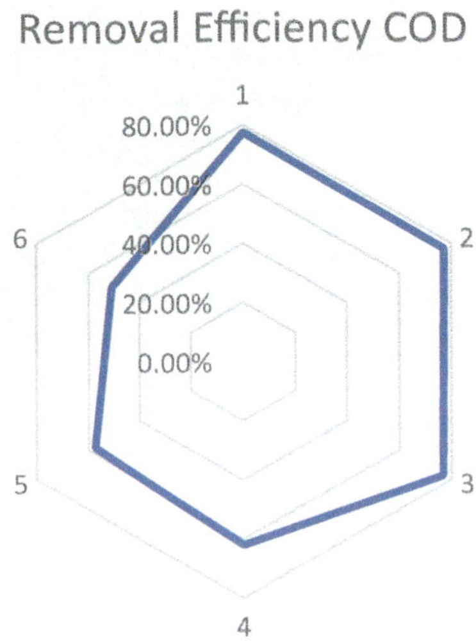

Fig. 6. Behavior of the BOD removal efficiency in the oxidation ditch

Likewise, it was verified that the maximum amount of BOD input was 142.55 mg/L and output after treatment with 31.7 mg/L, indicating that the contact time of the water with air should be increased, increasing the amount of entry of air and oxygen into the trench increasing agitation with the paddle, thus achieving air load values of 2.6 gr/day that guarantee greater oxidation of organic compounds, there must be an increase related to the operating flow of the 136% dissolved oxygen inside and outside of the oxidation ditch, as can be seen in Fig. 6. The behavior of the hydraulic conditions for the retention time in the cylindrical-based carousel-type Oxidation Ditch is 17 h. In comparison with the retention time used for the design, which is 18 h[30], which mentions hydraulic retention times from 17 h to 35 h, presenting a difference of 60 min.

Table 4. Confidence check for removal results

Variable	N	Mean	Std. Dev	Mín	Máx	G	P
COD inlet (mg/L)	6	203.8	98.2	102.4	298.8	1.03	1.000
COD outlet (mg/L)	6	58.07	10.16	48.00	69.30	1.11	1.000

With the results shown, it can be indicated that the level of confidence that was achieved in the removal reports reached 99.99%, achieved to the relationship of flow and dissolved oxygen regulated in the oxidation ditch, as seen in Table 4.

4 Conclusions

With a good operation and flow control monitoring, it can be concluded that 97% of the flow process time remained constant within the range of 0.48 to 0.55 m3/d, achieving that within the same oxidation ditch an Average pH of 8.01, being in good balance for the aerobic and anoxic processes, also managing to stabilize an average operating temperature of 17.9 °C within the carousel ditch. It was determined with a confidence level of 99.99% that the COD removal goes directly with the recirculation ratio of Qr/Qi = 0.752; managing to reduce organic matter up to a maximum of 77.76%, despite having atypical and high values, as was shown in the monitoring, achieving an amount of up to 142.55mg/L of BOD intake and obtaining a very high BOD output. Below the level of the environmental quality standard of Peru for category three that requires a minimum BOD of 31.7 mg/l to be used in irrigation and animal drinking,

It was also possible to demonstrate that it should be considered that the achieved range of dissolved oxygen in the ditch should be between 0.5 mg/l to 2.00 mg/l, which gave precision to ensure aerobic oxidation when operating with an average circulation of 0.519 m3/d and the shake times monitored show a good treatment capacity.

References

1. Edokpayi, J.N., Odiyo, J.O., Durowoju, O.S.: Impact of wastewater on surface water quality in developing countries: a case study of South Africa. Water quality **10**, 66561 (2017)
2. Sato, T., Qadir, M., Yamamoto, S., Endo, T., Zahoor, A.: Global, regional, and country level need for data on wastewater generation, treatment, and use. Agric. Water Manag. **130**, 1–13 (Dec. 2013). https://doi.org/10.1016/J.AGWAT.2013.08.007
3. Capodaglio, A.G.: Integrated, decentralized wastewater management for resource recovery in rural and peri-urban areas. Resources **6**(2), 22 (2017)
4. Bodík, I., Herdová, B., Drtil, M.: The use of upflow anaerobic filter and AnSBR for wastewater treatment at ambient temperature. Water Res. **36**(4), 1084–1088 (Feb. 2002). https://doi.org/10.1016/S0043-1354(01)00308-6
5. Van Veldhuizen, H., van Loosdrecht, M.C., Heijnen, J.J.: Modelling biological phosphorus and nitrogen removal in a full scale activated sludge process. Water Res**33**(16), 3459–3468 (1999)
6. Brdjanovic, D., van Loosdrecht, M.C.M., Versteeg, P., Hooijmans, C.M., Alaerts, G.J., Heijnen, J.J.: Modeling COD, N and P removal in a full-scale WWTP Haarlem Waarderpolder. Water Res. **34**(3), 846–858 (2000)
7. Cheng, C., et al.: Correlation of microbial community structure with pollutants removal, sludge reduction and sludge characteristics in micro-aerobic side-stream reactor coupled membrane bioreactors under different hydraulic retention times. Bioresour. Technol. **260**, 177–185 (2018)
8. Meijer, S.C.F., van Loosdrecht, M.C.M., Heijnen, J.J.: Metabolic modelling of full-scale biological nitrogen and phosphorus removing WWTP's. Water Res. **35**(11), 2711–2723 (2001)

9. Matko, T., Chew, J., Wenk, J., Chang, J., Hofman, J.: Computational fluid dynamics simulation of two-phase flow and dissolved oxygen in a wastewater treatment oxidation ditch. Process Saf. Environ. Prot. **145**, 340–353 (2021)
10. Terashima, Y., Ishikawa, M.: The kinetic analysis of BOD and nitrogen removal in an oxidation ditch. Water Sci. Technol. **17**(2–3), 291–302 (1985)
11. Chen, X., Fujiwara, T., Nakamachi, K., Kawaguchi, Y., Nishimura, F., Ohtoshi, K.: Evaluation of a novel oxidation ditch system with dual DO control technology for biological nutrient removal by mass balance analysis. Desalination **286**, 24–33 (2012)
12. Diaz, R., Mackey, B., Chadalavada, S., Heck, P., Goel, R.: Enhanced Bio-P removal: Past, present, and future–A comprehensive review. Chemosphere, 136518 (2022)
13. Wang, L.K., Shammas, N.K.: Single-sludge biological systems for nutrients removal. Advanced Biological Treatment Processes, 209–270 (2009)
14. Cardoso, B.J., Rodrigues, E., Gaspar, A.R., Gomes, Á.: Energy performance factors in wastewater treatment plants: A review. J. Clean. Prod. **322**, 129107 (2021)
15. Déléris, S., Paul, E., Audic, J.M., Roustan, M., Debellefontaine, H.: Effect of ozonation on activated sludge solubilization and mineralization. Ozone Sci. Eng. **22**(5), 473–486 (2000)
16. Nowak, O., Franz, A., Svardal, K., Müller, V., Kühn, V.: Parameter estimation for activated sludge models with the help of mass balances. Water Sci. Technol. **39**(4), 113–120 (1999)
17. Yang, M.: Dynamic mixing models and analysis of the mixing processes for an oxidation ditch and an aeration tank. Water Environ. Res. **94**(6), e10742 (2022)
18. Patziger, M., Kainz, H., Hunze, M., Józsa, J.: Influence of secondary settling tank performance on suspended solids mass balance in activated sludge systems. Water Res. **46**(7), 2415–2424 (2012)
19. Siddiqui, M.I., Rameez, H., Farooqi, I.H., Basheer, F.: Aeration control strategy design based on dissolved oxygen and redox potential profiles for nitrogen and phosphorus removal from sewage in a sequencing batch reactor. Journal of Water Process Engineering **50**, 103259 (2022)
20. Åmand, L., Olsson, G., Carlsson, B.: Aeration control–a review. Water Sci. Technol. **67**(11), 2374–2398 (2013)
21. Insel, G., Russell, D., Beck, B., Vanrolleghem, P.A.: Evaluation of nutrient removal performance for an ORBAL plant using the ASM2d model. In: Presentation, WEFTEC 2003 Conference, Los Angeles, California (2003)
22. Wang, X., et al.: Enhanced total nitrogen removal performance in a full scale Orbal oxidation ditch by a novel step aeration mode. Bioresour. Technol. **294**, 122228 (2019)
23. Bao, L., et al.: Progression of the silicate cathode materials used in lithium ion batteries. Chin. Sci. Bull. **58**, 575–584 (2013)
24. Stefani, M., Dobson, C.M.: Protein aggregation and aggregate toxicity: new insights into protein folding, misfolding diseases and biological evolution. J. Mol. Med. **81**, 678–699 (2003)
25. Huamán, M.M., Rosas, Y.W., Depaz, K.F.: Effect of Hydraulic and Organic Load, in the Removal of Biochemical Oxygen Demand in Wastewater using Biofilter with Vegetable Carbon, in High Andean Climate. In: IOP Conference Series: Earth and Environmental Science **973**(1) (2022). https://doi.org/10.1088/1755-1315/973/1/012002
26. Romero Rojas, J.A.: Tratamiento de aguas residuales, teoría y principios de diseño. Bogotá, CO, Escuela Colombiana de Ingenieros (2004)
27. Ibarra Suárez, D.V.: Eficiencia de la zanja de oxidación tipo carrusel de base cilíndrica en la planta piloto de agua residual ICS, en el centro experimental Tuyu Ruri–2021 (2022)
28. Federation, W.E., A. P. H. Association: Standard methods for the examination of water and wastewater. American Public Health Association (APHA): Washington, DC, USA (2005)
29. Rojas Vargas, R.: Zanjas de oxidación. In: Zanjas de oxidación 24 (1990)

30. Nieto, L.J., Ibarra, V.D.: PROPUESTA DE DISEÑO DE UNA PLANTA DE TRATAMIENTO DE AGUA RESIDUAL POR ZANJON DE OXIDACION PARA EL CASCO URBANO DEL MUNICIPIO DE VELEZ-SANTANDER. Universidad Católica de Colombia, Bogotá DC (2016)

Development of a Smart Framework to Continuously Monitor Constructed Wetland Efficiency

Henrique Pinho[1](✉), Manuel Barros[1], Pedro Matos[1], Carlos Ferreira[1], Pedro Neves[1], and Dina Mateus[2]

[1] Ci2 - Smart Cities Research Center, Instituto Politécnico de Tomar, Tomar, Portugal
hpinho@ipt.pt

[2] Techn & Art - Technology, Restoration and Arts Enhancement Center, Instituto Politécnico de Tomar, Tomar, Portugal

Abstract. Constructed Wetlands (CW) are Nature-Based Solutions for wastewater treatment. Although being a sustainable, eco-friendly, and low-cost technology, CW has some operative limitations because its efficiency regarding wastewater treatment depends on several uncontrolled variables. Besides many design options, the main mechanisms of pollutant removal depend on wastewater composition and environmental variables. Thus, optimization of CW operation requires the simultaneous knowledge of a large set of parameters. Moreover, control of wastewater flow rate, the single manipulable variable, requires nearly real-time knowledge of those parameters' variation. This work introduces the question of the effects of daily variation in the variables affecting CW efficiency and describes the EcoModZHC project, which aims the creation of a smart framework for real-time monitoring of CW. The EcoModZHC framework is based on water quality and environment variables sensors, supported by data acquisition and processing layers. Preliminary results demonstrated that daily variation of water quality is significant, causing deviations in the efficiency computation when traditional low-frequency sampling and laboratory analysis are applied. In addition, the EcoModZHC framework includes the control of a photovoltaic battery-assisted system, enabling the energetic autonomy of the complete CW system comprising the set of probes, pumps, and data acquisition and communication hardware.

Keywords: Sensor network · Water quality · Wastewater treatment

1 Introduction

Concepts like nature-based solutions (NBS) and circular economy are of increasing relevance for the strategic planning of resourceful, resilient, and circular cities because they can "build" an economy preserving the environment and promoting the well-being of the citizens [1, 2]. These concepts encompass concerns related to the preservation of the water cycle (ensuring quality demand and management), circular economy, energy production, ecosystem preservation, social and economic challenges, and the need to adapt to climate change [2].

J. Baeyens et al. (Eds.): ESRE 2023, pp. 14–26, 2025.
https://doi.org/10.1007/978-981-96-4345-5_2

Constructed wetlands (CW) are man-made systems that are designed to replicate the natural processes of wetlands to treat wastewater and are considered a cost-effective alternative to traditional wastewater treatment methods [3]. In this way, CW can be considered a NBS as they mimic the treatment processes of natural wetlands to efficiently treat wastewater [4]. They use the physical, chemical, and biological processes that occur in natural wetlands to remove pollutants and contaminants from wastewater, making it safer to discharge into the environment.

CW can be categorized into two main types based on their design and water flow, namely, free surface flow constructed (FSFCW) wetlands and subsurface flow constructed wetlands (SSFCW) [3]. Both types of constructed wetlands can be further subcategorized based on the type of vegetation and substrate used, the flow rate of the wastewater, and the specific pollutants being treated. The type of constructed wetland selected depends on various factors, such as the local climate, the quality and quantity of wastewater, and the goals of the treatment system.

Nature-based SSFCW typically consists of an impermeabilized basin or lagoon, designed to contain the wastewater and filler material, lined with an impermeable material, such as a plastic liner or concrete, to prevent the leakage of wastewater and to maintain the conditions necessary for the treatment process to occur.

SSFCW typically include the following components [3, 5]:

- Substrate: A permeable filler material, such as sand, gravel, or non-polluted solid waste, which provides a surface for the growth of plants and microorganisms.
- Macrophyte plants: Wetland vegetation, such as reeds, rushes, or cattails, that provide a habitat for microorganisms and help to remove pollutants from the wastewater.
- Microorganisms: Bacteria, fungi, and other microorganisms that break down pollutants in the wastewater through biological processes.
- Underdrain system: A system of pipes or a gravel layer that collects the treated wastewater and allows it to flow out of the CW.

All these components work together to create a self-sustaining system for the treatment of wastewater.

This work aims to present the development of the EcoModZHC project, which goal is to create a smart framework for real-time monitoring and analysis of the performance of CW treating wastewater. The following sections introduce the monitoring of CW (2), describe the prototype whose implementation is ongoing (3), describe the development of the sensors and data acquisition infrastructure (4), present preliminary results (5), describe a renewable system supporting the operation of the prototype (6), and state the mains conclusions and further work (7).

2 Monitored Parameters

The efficiency of wastewater treatment by CWs is usually based on the evaluation of physical, chemical, and biological parameters of water quality, as for conventional wastewater treatment plants [3, 6]. Physical parameters include Total Suspended Solids (TSS) and variants, Total Dissolved Solids (TDS) and variants, Electrical Conductivity (EC), and Temperature, among others. Chemical parameters usually refer to aggregate indicators, such as pH, Salinity, Total Nitrogen (TN), Total Phosphorous (TP), and Chemical Oxygen Demand (COD), among others. Dissolved Oxygen (DO), Nitrates, Nitrites, Phosphates, Chlorine, and metallic ions, among other ions and chemical elements, are also of common monitoring. Aggregated biological indicators such as Biochemical Oxygen Demand (BOD_5) or Total Coliforms, and specific microorganisms are examples of biological indicators whose monitoring is usual in CWs.

Table 1 presents the water quality parameters and frequency of sampling, reported by some recent works on CWs monitoring. All works referred to in Table 1 reported the use of laboratory methods after water samples collection. This procedure usually implies that the water sampling and consequent analytical evaluation are of low frequency. Almost all works report analysis of some indicator of organic compounds, mainly COD, and nitrogen compounds, through the main forms (ammonium, nitrate, and nitrite ions) but also TN, and phosphorous compounds, mainly through TP.

Table 1. Examples of water quality parameters evaluated in recent works on CWs and the sampling frequency.

Ref	Carbon/OC[1]	Nitrogen[2]	Phosphorous[3]	Other[4]	Frequency
[7]	COD	NH_4^+, NO_3^-, NO_2^-, TN		pH, DO	~ 1 week
[8]	COD	NH_4^+, NO_3^-, NO_2^-, TN		pH, DO, F, As, Cr	~ 1 day
[9]	COD	NH_4^+, NO_3^-, NO_2^-	TP		2 days
[10]	COD	NH_4^+, NO_3^-, TN	TP	PFAS	~ 1 day
[11]	TOC	NH_4^+, NO_3^-, NO_2^-, TN	PO_43-	pH, BPS, FPN, KTP, Br, SO_42-	1 week
[12]	COD, BOD_5	NH_4^+, NO_3^-, NO_2^-, TKN		DO, ORP, Dye, Color	1 week
[13]		TN	TP	TSS	1 week
[14]	COD	NH_4^+, NO_3^-, NO_2^-, TN	TP	pH, DO	3 days

[1]OC = Oxidable compounds; BOD_5 = Biochemical Oxygen Demand, 5 days, COD = Chemical Oxygen Demand; TOC = Total Organic Carbon
[2]TN = Total Nitrogen; TKN = Total Kjeldahl Nitrogen
[3]TP = Total Phosphorous
[4]BPS = bisphenol-group plasticizer bisphenol S; DO = Dissolved Oxygen, FPN = household-use insecticide fipronil, KTP = non-steroidal anti-inflammatory drug ketoprofen, ORP = Oxidation/Reduction Potential; PFAS = per- and poly-fluoroalkyl substances, TSS = Total Suspended Solids

3 Implementation of a Smart Framework Prototype

The EcoModZHC prototype can be described using the block diagram of Fig. 1. The three main sub-systems are:

1. A modular CW, made up of a set of interconnected wooden structures, in such a way as to be recyclable at the end of its life, makes a tank matrix that is easily scalable as required. More details are given by Pinho and Mateus [15].
2. A set of sensors and a water circulation system, composed of a set of environmental and water quality sensors and actuators to support the application (circulation pumps and valves), is responsible for acquiring the physical and chemical parameters at the system's input and output.
3. A monitoring and control system, which provides real-time information to the users of the installation, optimizes the process and guarantees the continuity of treated water quality.

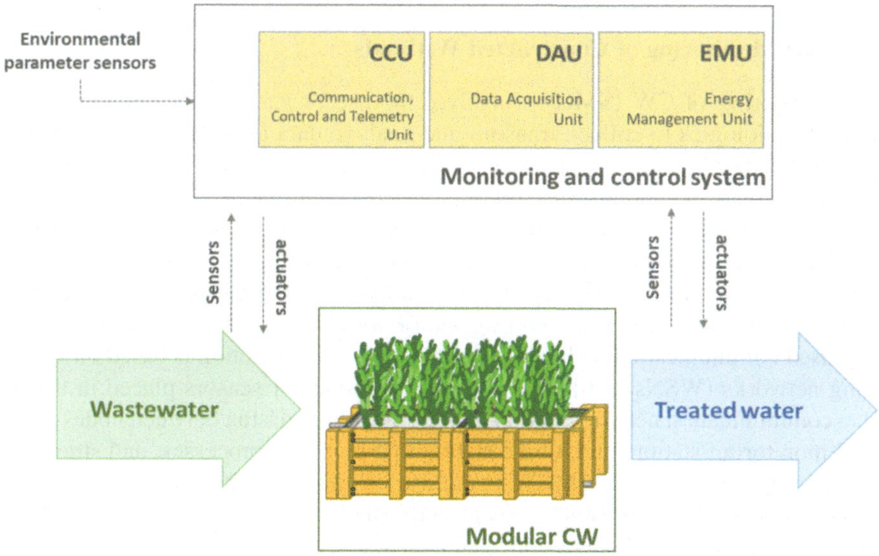

Fig. 1. EcoModZHC architecture diagram.

The flow of wastewater feeding the macrophyte plants is controlled by the entry and exit pumps, whose flow rate depends on the real-time measurements of the respective sensors. This infrastructure is divided into three subsystems: the Data Acquisition Unit (DAU), the Control and Telemetry Communication Unit (CCU), and the Energy Management Unit (EMU). The DAU unit will connect a set of environmental and water quality sensors and actuators to support the application. The interface is based on standard communication buses, to obtain a modular and configurable system to connect different types of sensors.

The CCU unit is the fundamental unit of the EcoModZHC system and consists of the development of a computational architecture based on the multitasking and real-time operating system, FreeRTOS, responsible for data acquisition, pre-processing, and transmission. This unit is divided into two subsystems: the control module and the telemetry module. The control module, supported by intelligent algorithms, controls the flow rate and the opening and closing of the water pumps. The telemetry module corresponds to a local communication infrastructure and a set of tools based on internet and cloud services for the generation and visualization of reports, the analysis of historical data, documentation, and trends, as well as the generation of statistics and alerts.

The Energy Management Unit, EMU, consists of the development of an alternative power system to guarantee the electric supply of the whole system and will be described in Sect. 6. The system will be based on batteries, recharged by photovoltaic panels, and should guarantee maximum autonomy to the EcoModZHC system, without loss of functionality, thus ensuring its integrity.

4 Development of the Sensor Network

4.1 Smart Monitoring of Constructed Wetlands

Smart monitoring of CW (SMCW) involves the use of various sensors and communication technologies to collect, transmit and analyze data about the performance and health of the wetland. The use of wireless sensing networks and advanced analytical tools enables real-time monitoring and improves decision-making, making CW a more sustainable and efficient solution for wastewater treatment. Sensors used are strategically placed within the wetland to measure various parameters such as water quality, water flow rate, and nutrient levels. The data collected from these sensors is then transmitted wirelessly to a central monitoring station for real-time analysis. The most commonly used communication technology for this type of application is based on wireless sensing networks (WSNs) [16]. WSNs consist of nodes or sensors placed in the wetland, a communication network that facilitates data transmission between nodes and the central monitoring station, and a base station that receives, processes, and stores data. Another important aspect of SMCW is the use of advanced analytical tools and software which enable real-time data analysis and visualization. Data science tools can be used to detect patterns in the data collected from the sensors, providing insights into the functioning of the CW, and identifying potential issues. In the following sections, the overall EcoModZHC system architecture, its components, and their interconnections are presented.

4.2 EcoModZHC Architecture Design Overview

In the absence of a standard reference architecture for the SMCW applications, a generic IoT 3-layer model architecture was adopted [17]. This worldwide accepted model defines a flexible reference architecture able to support the current features and future extensions of an IoT application. The IoT architecture implementation usually differs from application to application. The EcoModZHC IoT architecture system presented in Table 2 was designed taking into count the following design specifications: (i) ensure scalability and flexibility in future developments, in order to manage the increasing number of devices and services without compromising their performance; (ii) interoperability between system and applications, so that devices from different vendors can communicate to achieve common goals; (iii) distributive, to assure that data collected from different sources could be distributed, processed and analyzed by different entities and (iv) secure, to prevent unauthorized access.

Table 2. Architecture design of the monitoring system.

Layer	Component	Description
Perception Layer	Sensors	pH, electrical conductivity, temperature, turbidity, and ionic compounds; water level, environmental parameters such as temperature, humidity; and FreeRTOS (real-time operating system)
Network Layer	Data Transmission	I2C and RS485 Modbus (UART) Wifi (local Network) 3G (Internet) LoraWAN Broker MQTT (IoT protocol)
Application Layer	Data Storage	MongoDB (non-relational document database) MySQL (relational database) Cloud Storage (MRIPTA)
	MRIPTA - Web Data visualization	AdminLTE dashboard (Design Theme) HTML + CSS + JS and Laravel Framework ChartJS Webserver (plugin)
	Alert Services	SMS, SendGrid, Mailtrap (SMTP email)

4.3 IoT EcoModZHC Communication and Control Setup

The design of the EcoModZHC communication and control architecture is shown in Fig. 2. The core infrastructure is composed of the electronic components, sensors and actuators, the communication network, and the data acquisition and analysis platform (MRIPTA, https://mripta.ci2.ipt.pt/). The designed model will be described following the hierarchy defined in the three-layer model presented in Fig. 2: a) the perception Layer; b) the Network Layer; c) the Application Layer.

The Perception Layer. The Perception Layer is the lowest layer of the conventional IoT architecture and is responsible for collecting and processing data from various sensors and actuators that interact with the environment. Sensors play a crucial role in the SMCW, as they are responsible for collecting data about various parameters that are important for the performance of the wetland. A range of different electrochemical sensors was used in the input and output flow streams of this pilot-scale CW (Table 3). The feedback from the input and the control system's output stream is a fundamental component to satisfy the requirements of a fully automatic control-loop system. In this application, the control variable is given by the water flow stream at the input of the system. The sensors' feedback information is essential to control the performance stability of the system. A couple of ESP32 microcontrollers were used for the acquisition of water quality parameters from sensors such as pH, electrical conductivity, temperature, turbidity, and ionic compounds; and also for the acquisition of environmental parameters, such as the air temperature and humidity. The I2C and Modbus protocol were used to interface the water quality sensors and controllers, except for the CleanGrow® 200 device model, a closed system that uses the http protocol to provide additional water parameters information.

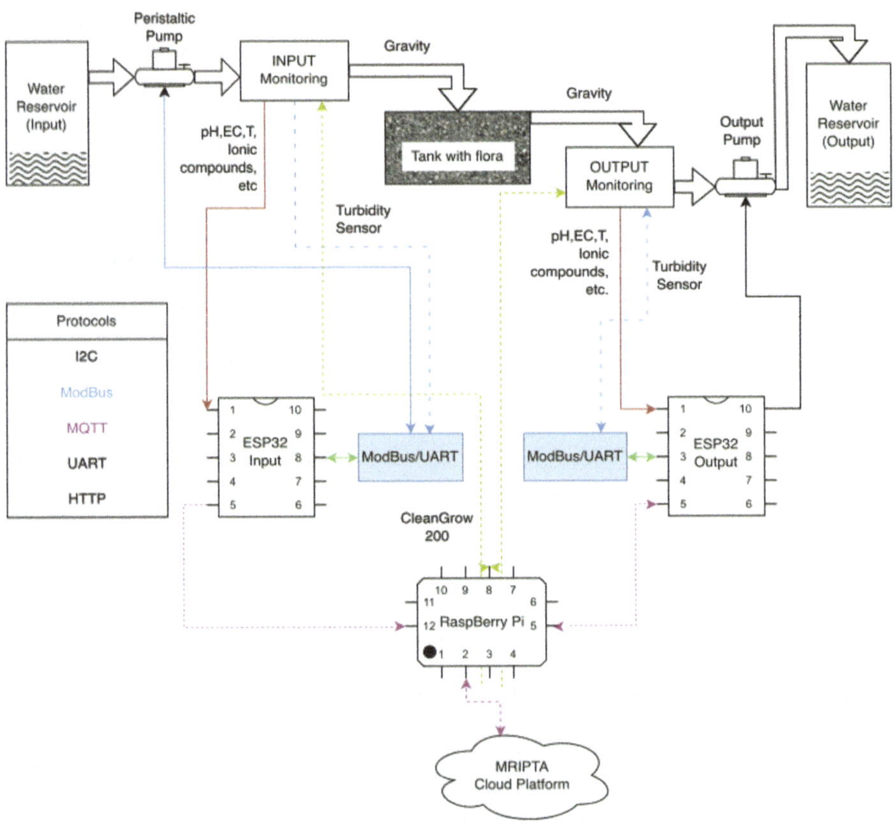

Fig. 2. The overall communication architecture.

The ESP32 microcontrollers also implement two additional functions: on the input side, the ESP is responsible for the management of a peristaltic pump, the device used to control the variable of the closed-loop systems. The ESP32 on the output side has the additional function of controlling the discharge of water from the tank at the system outlet. The control information can be generated locally, through a RPi microprocessor or generated by MRIPTA, the central monitoring platform in the cloud. The integrated system RPi/ESP32 at a local level was designed to incorporate both the LoRaWAN and Wifi technology for transmitting data from the sensors and control information to the actuators. The Wifi technology is primarily used to test and validate the system, but in the final version the LoRaWAN technology will be adopted, taking advantage of the worldwide collaborative network, called The Things Network, already installed in the Tomar region.

Table 3. Installed sensors and actuators.

Component	Model	Manufacturer
Nutrients analyzer	CG200	Clean Grow (UK)
pH probe	Ezo-PH	Atlas Scientific (USA)
Electrical conductivity probe	Ezo-EC	Atlas Scientific (USA)
Turbidity probe	PF-CAP-C-00146	Aqualabo (France)
Dissolved Oxygen probe	PF-CAP-C-00140	Aqualabo (France)
Peristaltic Pump	YZ15/YZ25	Lead Fluid Technology (China)

The Network Layer. This layer is responsible for processing real-time data between the perception (local) layer and the application layer, usually located remotely and achieved through telecommunications networks, local area network (LAN), and the Internet. The main component in this layer includes the RPi microcontroller that implements the functions of a gateway device of the network and provides connectivity, security, and data routing between the private application network and the public Internet. The publish-subscribe standard message protocol MQTT (Message Queue Telemetry Transport) was used for local communications because is well suited for bidirectional data transmission from many local devices to a single central monitoring station.

The Application Layer. Finally, the MRIPTA cloud-based platform was developed for storing, processing, and presenting data to the user and providing ways to interact with the application. The MRIPTA web framework was built using Laravel web technology (https://laravel.com) and provides a user-friendly interface for accessing, controlling, and presenting the application parameters. The software package was designed to implement analytic algorithms to analyze time-series data and provide system insights, enabling the EcoModZHC system to be more effective and efficient over time. The Dashboard is modular and configurable, with different views and windows, each of them will be able to manage users, produce reports, and define metrics and graphs. It also implements an alert management system that allows the creation of alert activation points and sends email notifications alerts via SMS or email, alerting users when problems arise.

5 Preliminary Results Obtained with the Sensor Network

The first experiments were carried out with pH, EC, and DO probes installed in one 1 × 1 CW prototype. Figure 3 shows a typical observed daily variation of pH, DO, and TDS from the EC probe.

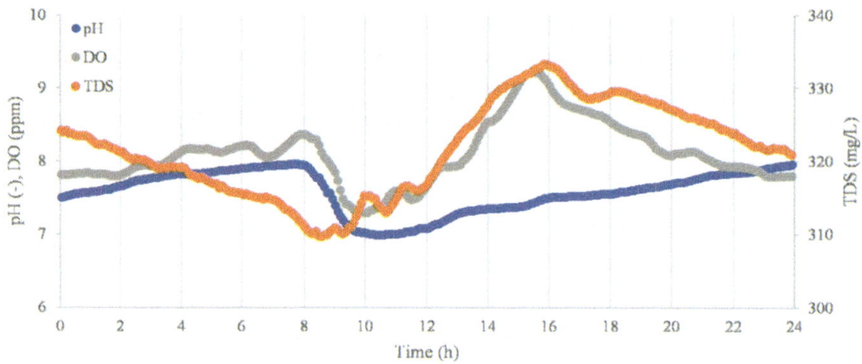

Fig. 3. Typical daily variation of pH, Dissolved Oxygen, and Total Dissolved Solids in the CW output. Data acquisition with 15 min frequency.

The variation of the 3 tested water parameters ranged from 6.99 to 7.96, with a standard deviation (SD) of 0.29, from 7.29 to 9.22 ppm with a SD of 0.43 ppm, and from 309.7 to 333.3 mg/L with a SD of 6.4 mg/L, respectively for pH, DO, and TDS. The daily averages are.

Considering, for example, that a single daily sampling was taken at 10 am, the observed results were 7.01 for pH, 7.29 ppm for DO, and 315.2 mg/L for TDS. These values have all a negative deviation from the daily average values: minus 7.6%, 10%, and 2%, respectively for pH, DO, and TDS.

For an input of wastewater with an almost steady flow rate and TDS average of 860 mg/L on that day, the over-evaluation of the dissolved solids remotion efficiency is less than 1% (62.6% considering the daily average TDS output value, and 63.3% considering the single sampling TDS evaluation). For this example, of day and water quality parameters, the deviation is acceptable. However, as previously referred, the optimization of CW operating can be done by manipulating the wastewater load into the system. Using this accurate knowledge of the daily variation of output water quality, it is possible to actuate the input flow rate, increasing the input load when the treatment efficiency is high, and decreasing the input load when the treatment efficiency is low. pH and DO are minor indicators of treated water quality, however, they have a significative impact on the three main contributions of CW for the removal of pollutants from wastewater: these variables affect the assimilation of nutrients by the plants, the degradation of pollutants by the microbiota community, and the adsorption and other mechanisms of pollutants sorption by the filling materials.

6 Energy Management

In this section, the Energy Management Unit, EMU, is described, mainly based on alternative energy sources to guarantee the electric supply of the whole system. Although an AC mains grid energy power supply is a possibility, the AC grid is not always available, depending on plant location. Additionally, climate change and pollution, impose the need for an energy transition to renewable energy sources.

As referred previously, the system will be based on batteries, recharged by photovoltaic panels, and should guarantee maximum autonomy to the EcoModZHC framework, without loss of functionality, thus ensuring its integrity by shortcoming light fluctuations due to weather variations and seasons.

Photovoltaic systems have the advantage of reducing atmospheric pollution, and avoiding CO_2 emissions. Each installed kilowatt (kW) of photovoltaic solar energy avoids the emission of 136 kg of CO_2 per year [18].

The identification of an approximation for power requirements (the system is not closed, so it is considering the scope for some evolutions in terms of sensors and signal processing) leads to Table 4.

Table 4. Sensors and actuators and respective consumption.

Component	Consumption
Sensors	7.5 W
Processors/communications	7.5 W
2 Peristaltic Pumps	2×60 W

The core consumption is composed of the peristaltic pumps, sensors, and data acquisition/communication systems, given a 135 W power need. For this power level, a 12 V system is adequate. The current is relatively low and so there is no need to escalate the voltage. Multiplying by 24 h gives 3240 Wh/day.

The sensors and actuators have different needs in terms of voltage. To maintain a low consumption, switching power converters are used to adapt the voltage levels. Considering an average efficiency of 85% in power conversion, the daily energy consumption became $3240/0.85 = 3811.8$ Wh.

The NASA database [19] or others like PVGIS [20] allows to take into account the critical month in which the solar hours are at a minimum, and so production is smaller. The sun hours are considered three for the present location and for the critical month, which is November. The peak power of the photovoltaic panels could now be calculated and is $3811.8/3 = 1270.6$ W.

To determine the battery capacity [Ah], the number of days that the battery will be discharging without receiving sufficient power from photovoltaic panels, due to the low solar irradiation, has to be taken into account. It is typically considered between three and ten days. Considering, in the present case a five day autonomy, a system voltage [V] of 12V and a DP (depth of discharge) of 0.8, the battery capacity should be: $(3811.8 \times 5)/(12 \times 0.8) = 1985$ Ah.

The system is represented in Fig. 4.

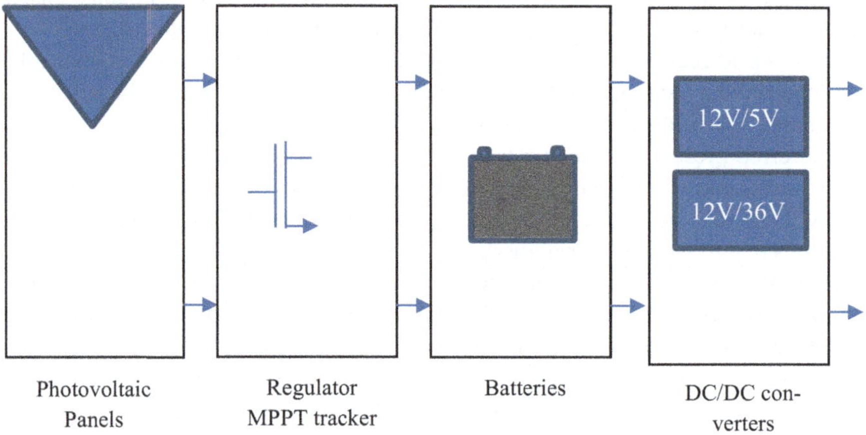

Fig. 4. Photovoltaic energy system block diagram.

7 Conclusions and Future Work

CW, like other nature-based systems, can contribute to sustainable development. CW, particularly, can have a direct role in the attainment of Sustainable Development Goal (SDG) six (Clean Water and Sanitation) by driving affordable water reclamation processes [21], and SDG seven (Affordable and Clean Energy) via the production of vegetal biomass for bioenergy applications [22]. However, optimal operation of CW, in which wastewater treatment efficiency depends on interconnected biological, chemical, and physical mechanisms, can be achieved through a better knowledge of the relationships between the variables underneath those mechanisms and environmental parameters.

Preliminary results obtained from a prototype of a smart framework based on water quality probes demonstrate that it can be observed deviations from low-frequency conventional laboratory analysis, and data obtained from continuous monitoring through a sensor network. Moreover, the framework under development can generate a set of data that can be a basis for future development of models based on data science tools, which can be applied to actively adapt the wastewater load to respond to variations in the treatment efficiency. Therefore, future work will be aimed to develop that sort of model.

Finally, the EcoModZHC framework shows that CW facilities can be designed as energy-autonomous systems through the use of renewable energy sources, such as photovoltaic systems.

Acknowledgments. This research was supported by the grant CENTRO-01-0145-FEDER-179932 (EcoModZHC project). The grant POCI-01-0145-FEDER-23314 is also acknowledged. The authors also thank the support of the BIOTEC.IPT Lab, Ci2, Lab.IPT and all supporting services of Politécnico Instituto de Tomar.

References

1. Pearlmutter, D., et al.: Enhancing the circular economy with nature-based solutions in the built urban environment: green building materials, systems and sites. Blue Green Syst. **2**(1), 46–72 (2020)
2. Calheiros, C.S.C., Castiglione, B., Palha, P.: Nature-based solutions for socially and environmentally responsible new cities: The contribution of green roofs. In: Stefanakis, A., Nikolaou, I. (eds.) Circular Economy and Sustainability, vol 2, pp 235–254. Elsevier, Amsterdam (2022)
3. Kadlec, R., Wallace, S.: Treatment Wetlands, 2nd edn. Taylor & Francis, Boca Raton (2009)
4. Cross, K., et al.: Nature-Based Solutions for Wastewater Treatment: A Series of Factsheets and Case Studies. IWA Publishing, London (2021)
5. Shukla, A., Parde, D., Gupta, V., Vijay, R., Kumar, R.: A review on effective design processes of constructed wetlands. Int. J. Environ. Sci. Technol. **19**, 12749–12774 (2022)
6. Ahmed, M., Al-Matouq, A., Khajah, M., Abdullah, H., Al-Ajeel, F.: Multi-criteria Assessment of Onsite Packaged Wastewater Treatment Systems. International Journal of Environmental Science and Development **13**(6), 292–297 (2022)
7. Zuo, J., et al.: Microbial community structure analyses and cultivable denitrifier isolation of Myriophyllum aquaticum constructed wetland under low C/N ratio. J. Environ. Sci. **127**, 30–41 (2023)
8. Wang, J.-F., Zhou, H.-Z., Tang, G.-H., Huang, J.-W., Liu, H., Cai, Z.-X., He, Z.-W., Zhu, H., Song, X.-S.: Reducing the inhibitive effect of fluorine and heavy metals on nitrate reduction by hydroxyapatite substrate in constructed wetlands. Journal of Hazardous Materials , 130692 (2023)
9. Kong, F., et al.: Influence of modified biochar supported sulfidation of nano-zero-valent-iron (S-nZVI/BC) on nitrate removal and greenhouse gas emission in constructed wetland. J. Environ. Sci. **125**, 568–581 (2023)
10. Ji, B., et al.: Curbing per- and polyfluoroalkyl substances (PFASs): First investigation in a constructed wetland-microbial fuel cell system. Water Res. **230**, 119530 (2023)
11. Ren, T., Perdana, M., Křížnecká, S., Sochacki, A., Vymazal, J.: Constructed wetlands for the treatment of household organic micropollutants with contrasting degradation behaviour: Partially-saturated systems as a performance all-rounder. Chemosphere **314**, 137645 (2023)
12. Benny, C., Chakraborty, S.: Dyeing wastewater treatment in horizontal-vertical constructed wetland using organic waste media. J. Environ. Manage. **331**, 117213 (2023)
13. Schwammberger, P., Tondera, K., Headley, T., Borne, K., Yule, C., Tindale, N.: Performance monitoring of constructed floating wetlands: Treating stormwater runoff during the construction phase of an urban residential development. Science of The Total Environmente **865**, 161107 (2023)
14. Li, C., et al.: Enhancement of organics and nutrient removal and microbial mechanism in vertical flow constructed wetland under a static magnetic field. J. Environ. Manage. **330**, 117192 (2023)
15. Pinho, H.J.O., Mateus, D.M.R.: Valorization of solid waste in subsurface flow constructed wetlands based on renewable modular structures: A contribution to a circular economy. In: Stefanakis, A., Nikolaou, I. (eds.) Circular Economy and Sustainability, vol. 2, pp 215–233, Elsevier, Amsterdam (2022)
16. Martínez, R., Vela, N., el Aatik, A., Murray, E., Roche, P., Navarro, J.M.: On the Use of an IoT Integrated System for Water Quality Monitoring and Management in Wastewater Treatment Plants. Water **12**(4), 1096 (2020). https://doi.org/10.3390/w12041096
17. Lombardi, M., Pascale, F., Santaniello, D.: Internet of Things: A General Overview between Architectures, Protocols and Applications. Information **12**(2), 87 (2021). https://doi.org/10.3390/info12020087

18. Serrano-Guerrero, X., Alvarez-Lozano, D., Romero, S.: Influence of local climate on the tilt and orientation angles in fixed flat surfaces to maximize the capture of solar irradiation: A case study in Cuenca-Ecuador, 2019 IEEE International Autumn Meeting on Power, Electronics and Computing (ROPEC), pp. 1–6 (2019)
19. NASA.: Power Data Access Viewer. Meteorological Data, available at: https://power.larc.nasa.gov/data-access-viewer/. Last Accessed 15 Feb 2023
20. European Commission.: Photovoltaic Geographical Information System. https://re.jrc.ec.europa.eu/pvg_tools/en/tools.html. Last Accessed 15 Feb 2023
21. Pinho, H., Mateus, D.: Sustainable production of reclaimed water by constructed wetlands for combined irrigation and microalgae cultivation applications. Hydrology **8**(1), 30 (2021)
22. Pinho, H., Mateus, D.: Bioenergy routes for valorizing constructed wetland vegetation: An overview. Ecol. Eng. **187**, 106867 (2023)

Review of Remediation Techniques in Agricultural Soils Contaminated by Heavy Metals in Latin America

M N Lara Ponce(✉), R P Mateo Coz, and J V Cornejo Tueros

Professional Academic School of Environmental Engineering, Continental University, Huancayo, Perú
70302281@continental.edu.pe

Abstract. The contamination of agricultural soils by heavy metals is an environmental problem, social and economic that is currently facing Latin America, in just the past 5 years has seen a reduction in agricultural production and performance of the soil. The review article aims to analyze the techniques of bioremediation most commonly used in Latin America in agricultural soils contaminated by heavy metals, through a method of review is a systematic, selecting certain publications with a minimum of 6 years of age and performing a detailed analysis of each country. When analyzing the studies, it is evidence that the majority of Latin American countries used the technique of phytoremediation because of the advantages it offers, it's easy to obtain, the efficiency in reducing the concentration of heavy metals and low cost. Also, it is concluded that in some countries of Latin America there are no publications relating to this topic, therefore, it is assumed that it is due to the lack of motivation and implementation of scientific and technological development by universities and states of each nationality.

Keywords: Bioremediation · heavy metals · phytoremediation · and contaminated agricultural soils

1 Introduction

For Latin America, agriculture accounts for between 5% and 18% of Product Gross Domestic in 20 countries [1] at the global level is exported to 23% of raw materials in agricultural and fishery products, in addition, has a projected growth of 28% for the 2028 [2]. According to the United Nations Organization for Food and Agriculture [3] the soil is defined as a non-renewable natural resource of high importance that has valuable functions in the development of the terrestrial ecosystem, ecological quality, maintenance of biodiversity, sustainability and evolution of all other resources. The fundamental need of the people is the power which is obtained through agricultural activities [4], however, in the agricultural sector, the land degradation is mostly due to anthropogenic activities, such as mining, poor disposal of wastes, agricultural practices, rapid urbanization, industrialization and oil spills. As a result, some consumers have shown the presence of heavy metals in the products, which has led to an economic

J. Baeyens et al. (Eds.): ESRE 2023, pp. 27–41, 2025.
https://doi.org/10.1007/978-981-96-4345-5_3

loss for the farmers who have invested in the export to a foreign market. The heavy metals (MP) are the main environmental pollutants of the type inorganic, hazardous by their toxicity, persistence and their cumulative capacity as Arsenic (As), Cadmium (Cd), Lead (Pb), Zinc (Zn), Chromium (Cr), Nickel (Ni), Mercury (Hg) and Copper (Cu), hereinafter referred to by its acronym [5]. The heavy metals are toxic substances are not biodegradable, and possess properties of biomagnification turning them into a threat to the resources such as water, soil, and air which are harmed because they have influnce in the agricultural sector and in the human health [6]. In high concentrations can impair human health, affecting the biological function of growth, and various organs, causing diseases that manifest in the long term, such as cancer, genetic diseases, and injuries of the nervous system [7].

In developing countries, the agricultural lands are contaminated by heavy metals, which constitutes a severe environmental problem due to the toxicity of these [6]. In Latin America, is extracted from the 44.6% of copper, and four of the five countries with the highest extraction of heavy metals such as: Peru (1° silver, 2° zinc, 3° copper and tin, 4° molybdenum, and lead, 5° gold), Chile (1° copper, lithium and iodine), Brazil (1° niobium, 2° iron, 3° bauxite, 5° tin) and Bolivia (3° antimony and 4° tin), mining is an activity that generates a great impact to the soil, after the agriculture moderna, which was regarded as one of the major sources of heavy metals, due to the use of fertilizers, particularly phosphorus and nitrogen, which harms the performance of the agricultural productivity, crop protection, agricultural systems, agronomy, ecology, genetics, and population dynamics.

To identify the types of metals and the amount of concentration in soils, must make an environmental monitoring, which should be guided by ethical and legal aspects such as regulatory law "Limits Maximum Permissible" (PML) which have been established in various countries, they have mostly been flexible with their parameters, allowing these contaminants affect significantly to the appeal of the soil in the agricultural sector. Such as in the countries of Peru, Cuba and Ecuador, which were found to be moderately contaminated with As, Pb, Cr, Zn and Ni in agricultural soils exceeding or being next to the LMP [7] [8] [9]. Through the years, it has been developing technologies to remediate soils contaminated by heavy metals, one of them is the environmental biotechnology implemented in the agricultural field [10]. Also, there have been new techniques and technologies that focus on the remediation of the resource from the soil through biological agents, nutrients and chemical compounds, such as bioremediation ex-situ and in-situ phytoremediation, biostimulation, among other [11]. However, it has limitations at the level of implementation, operational, and economic, in the majority of cases is also the lack of knowledge of use of these technologies in this sector. Therefore, it is important to continue research to find a viable solution that contributes to a better development of sustainable agriculture.

In this study we conducted a systematic review on the techniques of bioremediation used in agricultural soils that are contaminated by heavy metals in each country of Latin America. Our goal was to reflect the impact of heavy metals in agriculture and the pair analyze the techniques of bioremediation which help to reduce the concentrations of metals and enhance the development gap in Latin America.

2 Methodology

The present investigation was carried out on the basis of literature carried out in various countries of Latin America, in the databases of Google Scholar, Springer, Elsevier, Scielo, Researchgate, and in the repositories college, published between the years 2017 to 2022, with the use of key words in Spanish and English as: heavy metals, bioremediation, agricultural soils, Latin America, agriculture and pollution. It took into account the studies that had as objective to provide qualitative information and quantitative about the techniques of bioremediation in soils contaminated by heavy metals, with an emphasis (Zn, Cu, Cd, Cr, Hg, Pb, As). Subsequently developed inclusion criteria that allowed select and analyze the study of bioremediation in contaminated soils, only collected 17 scientific articles by the lack of research related to the application of the techniques of bioremediation in Latin America and for this reason, it was decided to supplement the study with 9 thesis undergraduate and graduate. Once you have completed the literature search, the found material was collected in an Excel file to be tabulated in order to know the relevant information that you provided to carefully study the parameters used in each technique.

3 Techniques of Bioremediation of Soil in the Agricultural Sector

To remediate contaminated sites, it is necessary to analyze and to examine the characteristics and types of pollutants, in addition to evaluate the structure of the site selection [4]. In order to choose the correct technique of remediation of soils with a focus on chemical, physical and biological, some of the most widely used technologies is the bioremediation due to their efficiency, transformation of hazardous contaminants and restore in its original form [12]. Figure 1 shows us the techniques of bioremediation depending on the site are categorized into ex situ or in situ. However, to choose the technique of bioremediation are considered some criteria such as the depth, degree of contamination, type of atmosphere, location, environmental policies, cost and performance criteria, all of which have advantages and disadvantages [13].

3.1 Techniques of Bioremediation

To be able to remediate contaminated sites it is necessary to know the advantages and disadvantages of each bioremediation technique, that's why a deep review of each technique was carried out to know which would be the most indicated according to studies already carried out in Latin America either for its efficiency, time and cost. As shown in Table 1.

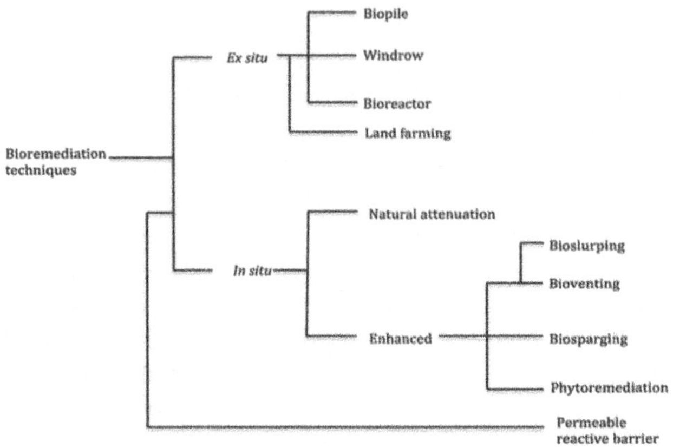

Fig. 1. Techniques of bioremediation according to their type of selection [13].

There are several techniques within the phytoremediation as: phytoextraction or phytoaccumulation, phytodegradation, phytovolatilization, phytorestauration and phytostabilization.

Figure 2 shows the types of phytoremediation plants and micro-organisms that are associated to a better role in the recovery of contaminated soils, being based on a natural process through the use of native plant species associated with the degradation of pollutants being often considered as an alternative to the technologies of remediation for being an activity that is economically sustainable, efficient and respectful with the environment.

4 Heavy Metals in Agricultural Soils

Heavy metals are pollutants that are mostly found associated with the soil, in some cases with natural origin when it is derived from parent material or activities carried on by the man [29]. In addition, agricultural activities can contaminate the soil in the long term, due to the use of fertilizers and agrochemicals, in which compound are heavy metals; also the long-term use of compost and sewage, untreated, can lead to increased levels of heavy metals in the soil, throwing values in excess of those permitted in the national and international legislation [30]. Therefore, in Table 2, we show the most common heavy metals found in agricultural soil and the impact they generate on it.

Table 1 Review of Techniques of Bioremediation in Latin America.

	Description	Advantages	Disadvantages
Phytoremediation	Used through the use of plants, which perform catches of toxic substances, due to the fact that their roots have high absorption capacity [14]. There are several techniques within the phytoremediation as: phytoextraction or phytoaccumulation, phytodegradation, phytovolatilization, phytorestauration and phytostabilization. Figure 2 [15]	Respects the environment, is the efficient long term and less expensive [16]. High adsorption capacity of heavy metals, covering large surfaces, it is reusable and easy to implement. Reduce approximately 15.3 to 84.6% heavy metals [17]. Technology sustainable and cost effective	To have a good efficiency must be combined with other treatments. Lower efficiency in soils heterogeneous [16]. It has some limitations to certain metals and other inorganic compounds in the soil
Organic amendments	This process focuses on the use of the product of plant or animal origin, whose principal objective is to maintain and increase the con centration of organic matter in the soil and the quality of their physical-chemical properties, to more easily absorb the nutrients [18]	Fast and active microbial growth. [18] Used as sinks aimed at reducing the bioavailability of the metal	Hinder the process of bioremediation when not used correctly, the amendments, and the excess of nutrients is detrimental to the microbial community [19]
Bioaugmentation	Technique in using microorganisms grown with the goal of biodegrade the contaminants from soil or water [20]	Improves the degradation and rate of transformation of xenobiotics by means of insertion of microorganisms [21]	Bioavailability of the contaminant [22]
Biopila	Is the piling up of contaminated soil, are sometimes deployed with other techniques, such as amendments with nutrients and aeration. [13]	Is achieved by an effective biodegradation. It is effective on sites ends contaminated. The flexibility of the biopila allows you to shorten the time of remediation	You have to take a control of the temperature and aeration. In comparison to other techniques occupy a large space. High cost of maintenance
Rows	Consists of the flip newspaper of the contaminated soil, by means of activities of degradation of the hydrocarbon clastic bacteria. [13]	Best technique for soils contaminated with hydrocarbons	It is not so recommended for soils contaminated with volatile toxic

(continued)

Table 1 (*continued*)

	Description	Advantages	Disadvantages
Bioreactor	Technology that consists of 4 parts, management of soil pollutants, the installation of the handling and disposal of treated soils and battery of the bioreactor	More efficient in soil contaminated by recalcitrant in controlled environmental conditions. Increased contact between microorganisms [23]	More expensive than other techniques of bioremediation [23]
Bioventilation	Is a mechanism for airing the horizon of the soil and stimulate the bacterial decomposition in the soil profile	They are efficient in shallow depths [24]	The restoration of contaminated sites is longterm. It is most commonly used for groundwater
Biosorption	Refers to the uptake of metals that leads out to a biomass-complete (dead or alive), through mechanisms physicochemical as adsorption or ion exchange	They are efficient in the generation of alkalinity, effective for the removal of metals multimetallic [25]	You can only treat the metals at low concentrations. It is necessary to feed the flows under conditions of physiologically permissible [26]
Bioaspersion	This technique consists of injecting air into the subsoil with the goal to stimulate microbial activities	Most commonly used in treatments contaminated by products of diesel and kerosene in aquifers	Predict the direction of the air flow
Permeable reactive barrier	Is a technique for the remediation of organic pollutants and can also be applied to the neutralization of the pH. [27]	Treatment in-situ high potential. Overall efficiency for the removal of metals, and more than 70% of the sulfate	High cost of maintenance. Useful life of the adsorbent materials
Mycoremediation	Based on the use of ability of fungal strains that are responsible for secreting enzymes that manage to digest substrates, so that can reduce a high level of pollution [28]	It is versatile, can be performed in situ or ex situ. Requires little space. Low cost	It can take quite a long time to remediate the soil. If there are various pollutants, it is difficult to absorption

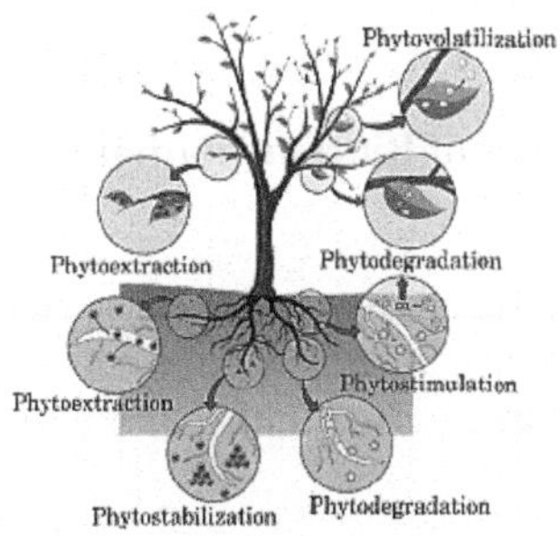

Fig. 2. Types of phytoremediation [15].

Table 2 Impact generated by the heavy metals in soils agricultural

Heavy metals	Properties pollutants	Impact
Zn	Is a metal that is chemically active	Considered risky for plants, has a high content of phosphorus and silicon
Cu	Difficult to degrade	Exposure by ingestion, absorption and inhalation of particles that come from the soil
Cd	Soluble in the soil solution, is manifested as chlorosis	Considered to be dangerous in a state of low concentrations, affects organisms in terrestrial and aquatic
Cr	Cr (0), Cr (III) trivalent and Crthe Cr (VI) are the species of chromium, more stable and frequent	Potential Risk to the agricultural sector
Hg	Compound of mercury Hg (Hg + 1), inorganic mercury (Hg + 1) and organic compounds of mercury (Hg + 2)	Toxic substance, can be found in three forms (elemental, inorganic and organic)
Pb	Acid, ammunition, metal products. It Is highly toxic in natural sources	Affect the tissues physiologically active plant
As	Difficult to degrade	Cause reduced growth and yield of the plant

Source[31–33]

5 Result

5.1 Compilation of Techniques of Bioremediation Used in the Agricultural Sector in Latin America

In the following Table 3 we conducted a systematic review, which came to analyze the percentage of efficiency for each pollutant and the time of absorption according to the technique of bioremediation that are used in the agricultural sector in various countries of Latin America [34–56].

5.2 Countries of Latin America, that Examined the Techniques of Bioremediation

Figure 3 indicates that Peru is the country with the highest research in the use of techniques of bioremediation, being the main technique phytoremediation with 60%, due to your great advantage and feasibility of implementation level, operational, and economic. Other techniques used are organic amendments, which were implemented in the largest quantity in Chile with 25%. On the other hand, takes into account the many other techniques that contribute to the treatment of contaminated soils.

Table 3 Review of the tremoval techniques of bioremediation and its efficiency in the agricultural sector in Latin America

Technique of bioremediation	% Efficiency	absorpti onTime	Count ry,	References or authors
Use of remediation physicochemical	99.48% for chromium, 81.48% cadmium, and 5.9% lead	*	Peru and Brazil	[34]
Organic Amendments	Biochar 80% Cal 60%	30-day	Colombia	[35]
Amendments organic	Was achieved to decrease the content of Pb removable with EDTA between 27% and 62%	60-day	Peru	[36]
Mycoremediation	Index of tolerance of 0.81 to an exhibition of300 mg/kg of Cd	13 days of incubation	Colombia	[37]

(*continued*)

Table 3 (*continued*)

Phytoremediation	Value of saturation of absorption of Cu 75%, Cd and Pb 50%, and 75%	to 120 days	in Peru	[38]
Phytoremediation Phytoremediation + Organic Amendments	Pb at the level of root 16,47% Cadmium (95 days) 0,94 %	125 days	Peru	[38]
Amendments	Removal for cadmium are 11% and to lead 9.951%	64 days	Peru	[39]
- organic (waste of wine-making)	Increase of absorption from 15 to 75%	in 30-day	Chile	[40]
Phytoremediation	Efficiency of 99%	*	Chile	[41]
organic Amendments (phosphate, di-ammonium FD, iron sulfate, SH, phosphate, ferrous FF, calcium peroxide PC and matters of organic MO)	30% Decrease of arsenic in a soil with amendments vs a land without	75 and 85 days	Chile	[41]
Phytoremediation	As, Cd, Cr, Cu, Fe,Hg, Mn, Se, Pb, and Zn	*	Mexico	[42]
Phytoremediation	Removal of cadmium to the 56,03 %. and lead to the 96,23 %	120 to 190 days in	Colombia	[43]
Biosorption	Removal of cadmium to the 98,63 %, 96,06 % and 88,93 %	11 days	Venezuela	[44]
Mycoremediation	Removal of Hg (57%)	*	Colombia	[45]
Phytoremediation	Removal of the 58% of Cd	*	Mexico	[46]
Phytoremediation (phytoextraction),	Cu, Co, Zn and Pb.	*	Cuba	[47]
Phytoremediation	Cd, Cr, Cu, Ni, Pb and Zn	1080 days	Argentina	[48]
Phytoremediation	Removal of the 18-28% Cr, Cu, Ni, and Zn	> 6 months	Argentina	[49]

(*continued*)

Table 3 *(continued)*

Phytoremediation (Helianthus annus)	Removal of the Cd 100% with 6 floors of H. annuus	40 days	Peru	[49]
Phytoremediation (alfafa, swiss chard, and amaranth)	Removal of 90% of Pb	60 days	Ecuador	[50]
Phytoremediation (phytoextraction, phytostimulation and phytoaccumulation)	Pb, Cd, As, Ag, Cu, Ni, Zn Cd (BCF>1, TF<1)	*	Peru	[51]
Phytoremediation (phytostabilization)	BCF= Factor bioconcetration TF= Factor translocation	*	Peru	[52]
Bioremediation, phytoremediation	Cu. Ni, Cr, Pb, Cd	*	Colombia	[53]
Bioremediation	Removal of Zn in a 50%	3 days of incubation	Ecuador	[54]
Phytoremediation	Removal 73% for Ace, 68% for Cu and 78% for Pb.	*	Chile	[55]
Phytoremediation	Cu 92.9% > Faith 85.7% > Pb 75.0% > 64.3% Zn	6 months	Mexico	[56

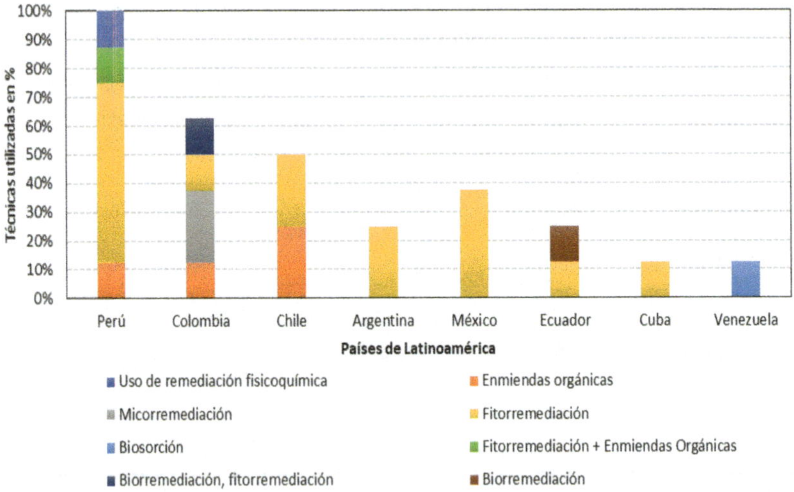

Fig. 3. Techniques of bioremediation most commonly used in Latin America.

6 Conclusion

This research details the use of the various technologies of bioremediation in the agricultural sector in Latin America, to mitigate and reduce the concentrations of inorganic contaminants in the soil, the increase in the concentrations of metals such as As, Cd, Cu, Pb, Zn, Ni, and Cr is reflected day-to-day in the present. Therefore, we analyzed in detail the research of various Latin American countries were analyzed in detail as seen in Table 3, concluding that the most widely used technique is the "phytoremediation" for their great advantage in various aspects such as efficiency, cost, variations, and ease of obtaining species vegetative to the absorption of heavy metals, it also takes into account the combination of types of biore-mediation to reduce the inorganic contaminants in less time, but with higher costs than the traditional. Also, it was found that the country which is carrying out further research in all of Latin America on the subject, published in articles and thesis Peru, and it showed that countries such as Uruguay, Honduras, Puerto Rico, Haiti and others do not have or are not partakers of research. So, it can be inferred that due to the lack of education and research in institutions of higher education. Therefore, this information denotes the interest and importance of the use-value of these various technologies of bioremediation in the agricultural sector that allow for respect environmental legislation in your field, while avoiding further contamination and risking our resources and health.

References

1. World bank IBRD-IDA.: Agricultural systems and food of Latin America and the Caribbean are ready for a profound transformation [Internet] (2020) [cited 2023 Feb 22]. Available from: https://www.bancomundial.org/es/news/press-release/2020/11/12/agriculture-food-systems-latin-america-caribbean-changes
2. Organization of the United Nations Food and Agriculture O.: OECD-FAO Agricultural Outlook 2020–2029 [Internet], 352 (2020) [cited 2023 Feb 22]. Available from: https://doi.org/10.1787/a0848ac0-es
3. Organization of the United Nations Food and Agriculture F.: State of the World's Soil Resource [Internet] **I**, 1–92 (2015). Available from: http://www.fao.org/3/a-i5126s.pdf
4. Yaashikaa, P.R., Kumarm, P.S.: Bioremediation of hazardous pollutants from agricultural soils: A sustainable approach for waste management towards urban sustainability. Environ Pollut [Internet], 312: 120031 (2022 Nov 1); [cited 2023 Feb 21]. Available from: https://www.sciencedirect.com/science/article/abs/pii/S0269749122012453
5. Soto-Benavente, M., Rodriguez-Tasteless, L., Olivera, M., Arostegui Sanchez, V., Hill, Nano-C., Garate Quispe. J., et al.: Health risks of heavy metals in agricultural products grown in areas abandoned by the gold mining in the peruvian Amazon. Sci Agropecu [Internet] **11**(1), 49–59 (2020 Mar 1) [cited 2023 Feb 21]. Available from: http://www.scielo.org.pe/sci-elo.php?script=sci_arttext&pid=S2077-99172020000100049&lng=es&nrm=iso&tlng=es
6. Lemessa, F., Simane, B., Seyoum, A., Gebresenbet, G.: Analysis of the concentration of heavy metals in soil, vegetables and water around the bole Lemi industry park, Ethiopia. Heliyon [Internet] **8**(12), e12429 (2022 Dec 1) [cited 2023 Feb 21]. Available from: https://www.sciencedirect.com/science/article/pii/S2405844022037173
7. Guzman-Morales, R., Amber, Cross-Peace Or, Valdés-Carmenate R.: Effects of pollution by heavy metals in a soil with agricultural use. Rev Technical Sciences Agropecu [Internet] **28**(1),

1–9 (2019). Available from: https://open.to/a/UZ5jD%0Ahttps://search.ebscohost.com/login.aspx?direct=true&db=fua&AN=135117310&lang=es&site=ehost-live

8. Organization of the Nations FU for the A and the A.: soil Pollution: A hidden reality, 144 (2019) [cited 2023 Feb 21]. Available from: https://www.fao.org/3/i9183es/i9183es.pdf
9. Romero-González BN.: Pollution by heavy metals in Food in Ecuador Meta-Analysis [Internet]. [Guayaquil]: University Guayaquil (2020) [cited 2023 Feb 22]. Available from: http://repositorio.ug.edu.ec/bitstream/redug/48744/1/TESIStoGROUND.PDF
10. Fajardo Gomez, C.A.: Bioremediation of heavy metals [Internet]. Books National University Open and Distance learning (2017) [cited 2023 Feb 21]. Available from: https://hemeroteca.unad.edu.co/index.php/book/article/view/3992/3977
11. Geaneth Valenzuela, B. I., Galindo, L., Mantilla, D., Moncada, D.S., Orjuela, E., Roman K., et al. Techniques of Bioremediation of Soil Contaminated by Hydrocarbons for use in the Municipality of Tibú, Norte De Santander **51**(51), 107–118 (2021) [cited 2023 Feb 22]. Available from: https://dialnet.unirioja.es/servlet/articulo?codigo=8234914
12. Nava-Reyna, E., Medrano-Macías, J.: Arsenic occurrence in the environment: Current situation of the Comarca Lagunera in northern Mexico and bioremediation approaches. J Agric Food Res [Internet] **10**, 100379 (2022 Dec 1) [cited 2023 Feb 21]. Available from: https://www.sciencedirect.com/science/article/pii/S2666154322001120
13. Azubuike, C.C., Chikere, C.B., Okpokwasili, G.C.: Bioremediation techniques–classification based on site of application: principles, advantages, limitations and prospects. World J Microbiol Biotechnol [Internet] **32**(11), 1–18 (2016) [cited 2023 Feb 22]. Available from: https://www.ncbi.nlm.nih.gov/pmc/articles/PMC5026719/
14. Mendarte-Alquisira, C., Alarcón, A., Ferrera-Cerrato, R.: Phytoremediation: An Alternative biotechnology to retrieve soil contaminated with DDT. A review. TIP Rev Spec Sciences Chemical-Biological [Internet] **24** (2021 Jun 19) [cited 2023 Feb 22]. Available from: https://redalyc.org/journal/432/43271644020/html/
15. Marrero, J., Amores, I., Coto, O.: Phytoremediation, a technology that involves plants and microorganisms on environmental sanitation. ICIDCA [Internet] **11** (2012 Sep 3) [cited 2023 Feb 21]. Available from: http://www.redalyc.org/articulo.oa?id=223124988007
16. Ahmed, S.F., Kumar, P.S., Rozbu, M.R., Chowdhury, T.A., Nuzhat, S., Ralph, N., et al.: Heavy metal toxicity, sources, and remediation techniques for contaminated water and soil. Environ Technol Innov [Internet]. **25**, 102114 (2022 Feb 1) [cited 2023 Feb 21]. Available from: https://linkinghub.elsevier.com/retrieve/pii/S235218642100746X
17. Chugh, M., Kumar, L., Shah, M.P., Bharadvaja, N.: Algal Bioremediation of heavy metals: An insight into removal mechanisms, recovery of by-products, challenges, and future opportunities. Energy Nexus [Internet] **7**, 100129 (2022 Sep) [cited 2023 Feb 21]. Available from: https://www.sciencedirect.com/science/article/pii/S2772427122000845
18. Huaraca, J., Perez, L., Bustinza, L., Pampa, N.: organic Amendments on the immobilization of cadmium in agricultural and contaminated soils: a review. Inf Technology [Internet] **31**, 14 (2020 Aug) [cited 2023 Feb 21]. Available from: https://doi.org/10.4067/S07180764202000 0400139
19. Mansfield, I.M.S., Isa, Y.M.: Bioremediation of acid mine drainage – Review. Alexandria Eng J [Internet] **65**, 1047–1075 (2023 Feb 15) [cited 2023 Feb 21]. Available from: https://www.sciencedirect.com/science/article/pii/S1110016822006470
20. Romero-Silva, R., Sanchez-Reyes, A., Diaz-Rodriguez, et al.: Bioremediation of soils contaminated with petroleum solid wastes and drill cuttings by Pleurotus sp. under different treatment scales. SN Appl Sci [Internet] **1**(10), 1–9 (2019 Oct 1) [cited 2023 Feb 22]. Available from: https://doi.org/10.1007/s42452-019-1236-3
21. Sánchez-Jiménez, A., Medrano-Roldán, H., Erika, K., Chavez-Avilés, M.N., Valiente-Banuet, J.I., Fierros-Rosemary, G.: Bio and phytoremediation: plants and microbes to the rescue of

heavy metal polluted soils. SN Appl Sci [Internet] **4**(2), 1–14 (2022) Feb 1 [cited 2023 Feb 21]. Available from: https://doi.org/10.1007/s42452-021-04911-y

22. Gonçalves, C.R., Delabona, P. da S.: Strategies for bioremediation of pesticides: challenges and perspectives of the Brazilian scenario for global application – A review. Environ Adv [Internet] 8 (2022 Jul 1) [cited 2023 Feb 21]. Available from: https://www.sciencedirect.com/science/article/pii/S2666765722000552
23. Robles-González, I.V., Fava, F., Poggi-Varaldo, H.M.: A review on slurry bioreactors for bioremediation of soils and sediments. Microb Cell Fact [Internet] **7**(1),1–16 (2008 Feb 29) [cited 2023 Feb 21]. Available from: https://doi.org/10.1186/1475-2859-7-5
24. Al-Weshah, R.A., Yihdego, Y.: Multi-criteria decision approach for evaluation, ranking, and selection of remediation options: case of polluted groundwater, Kuwait. Environ Sci Pollut Res [Internet] **25**(36), 36039–36045 (2018 Dec 1) [cited 2023 Feb 21]. Available from: https://doi.org/10.1007/s11356-018-3723-2
25. Jayapal, A., Chatterjee, T., Sahariah, B.P.: Bioremediation techniques for the treatment of mine tailings: A review. Soil Ecol Lett 2023 52 [Internet] **5**(2), 1–15 (2022 Dec 29) [cited 2023 Feb 21]. Available from: https://doi.org/10.1007/s42832-022-0149-z
26. Cañizares-villanueva, R.: Biosorption of heavy metals by the use of microbial biomass. Rev Latinoam Microbiol [Internet] (2000) [cited 2023 Feb 21]. Available from: https://www.medigraphic.com/pdfs/lamicro/mi-2000/mi003f.pdf
27. Farage, R.M.P., Quina, M.J., Gando-Ferreira, L., Silva, C.M., de Souza J.J.L.L., Towers, C.W.M.E.: Kraft pulp mill dregs and grits as permeable reactive barrier for removal of copper and sulfate in acid mine drainage. Sci Rep [Internet] **10**(1), 1–10 (2020 Dec 1) [cited 2023 Feb 21]. Available from: https://link.springer.com/articles/s41598-020-60780-2
28. Bedoya Cyrus, C.C., Hernando-Estupiñan, L.: In vitro Evaluation of the ability biorremediadora of filamentous fungi on crude oil, 22 (2018 Jan) [cited 2023 Feb 22]. Available from: http://www.scielo.org.co/pdf/nova/v16n30/1794-2470-nova-16-30-00037.pdf
29. Mahecha-Polished, J.D., Trujillo-González, J.M., Torres-Mora, M.A.: Analysis of Studies in Heavy Metals in Agricultural Areas of Colombia Análise de Estudos em Metais Heavy'em Agricultural Areas ee Colombia. Orinoco Basin [Internet] 12 (2017) [cited 2023 Feb 21]. Available from: http://www.scielo.org.co/scielo.php?script=sci_abstract&pid=S0121370920170003000083&lng=en
30. Mendoza-Escalona, B., Torres-Rodríguez, D., Marked, L.M., Gómez, C., Estanga-Barrios, M., Garcia-Orellana, y., et al.: Concentration of heavy metals in agricultural soils under different tillage systems. Technology [Internet] **24**(51), 4–15 (2021 Feb 26) [cited 2023 Feb 22]. Available from: http://www.scielo.org.co/scielo.php?script=sci_arttext&pid=S012377992021000200004&lng=en&nrm=iso&tlng=es
31. Alsafran, M., Saleem, M.H., Al Jabri, H., Rizwan, M., Usman, K.: Principles and Applicability of Integrated Remediation Strategies for Heavy Metal Removal/Recovery from Contaminated Environments. J Plant Growth Regul. [Internet] 22 (2022). Available from: https://doi.org/10.1007/s00344-022-10803-1
32. Puga, S., Sosa, M., Lebgue, T., Quintana, C., Fields, A.: Pollution by heavy metals in the soil caused by the industry. Ecol Appl [Internet] **5**(1), 72006 [cited 2023 Feb 21]. Available from: http://www.scielo.org.pe/pdf/ecol/v5n1-2/a20v5n1-2.pdf
33. Gardens Galan, E., Romero Baena, A.: Pollution of Soils by Heavy Metals. Rev Soc Spanish Ore **13** (2008)
34. Baptist-Guzman, J., Gomez-Morales, R., Asmat-Campos, D., Checca, N.R.: Influence of the Alcoholic/Ethanolic Extract of Mangifera indica Residues on the Green Synthesis of FeO Nanoparticles and Their Application for the Remediation of Agricultural Soils. Mol [Internet] **26** (24), 7633 (2021 Dec 16) [cited 2023 Feb 21]. Available from: https://www.mdpi.com/1420-3049/26/24/7633/htm

35. Becerra-Agudelo, E., Lopez, J.E., Betancur-Garcia, H., et al.: Assessment of the application of two amendments (lime and biochar) on the acidification and bioavailability of Ni in Ni-contaminated agricultural soils of northern Colombia. Heliyon. **8**(8), 9 (2022 Aug 1)
36. Hair-Torres, R.J., Romero-Longwell, J.R., Valdiviezo-Gonzales, L., et al.: Biochar derived from excreta pig with the ability to decrease the availability of Pb in agricultural soils contaminated. Sci Agropecu [Internet] **12**(4), 461–470 (2021 Dec 1) [cited 2023 Feb 21]. Available from: https://www.scielo.org.pe/scielo.php?script=sci_arttext&pid=S2077-99172021000400461&lng=es&nrm=iso&tlng=es
37. War Sierra, B.E., Arteaga-Figueroa, L.A., Sierra-Pelaéz, S., Alvarez, J.C.: Talaromyces santanderensis: A New Cadmium-Tolerant Fungus from Cocoa Soils in Colombia. J Fungi [Internet] **8**(10), 1042 (2022 Oct 1) [cited 2023 Feb 21]. Available from: https://www.mdpi.com/2309-608X/8/10/1042/htm
38. Huaranga Moreno, F., Méndez García, E., Quilcat Leon, V., et al.: Quantification of Cu, Pb, s and Cd absorbed by the sunflower Helianthus annus L. (Asteraceae), which are present in agricultural soils contaminated by mining wastes. Arnaldoa [Internet] **29**(1), 119–136 (2022) [cited 2023 Feb 21]. Available from: http://www.scielo.org.pe/scielo.php?script=sci_arttext&pid=S2413329920220001001l9&lng=es&nrm=iso&tlng=es
39. Grandez Argomeda, M.G.A.: Removal of cadmium and lead in soils on the banks of the River Mantaro Junin, by phytoremediation with sunflower (Helianthus annus) and maize (Zea Mays) using amendments [Internet]. César Vallejo university (2017) [cited 2023 Feb 22]. Available from: https://repositorio.ucv.edu.pe/bitstream/handle/20.500.12692/3537/Grandez_AMGA.pdf?sequence=1&isAllowed=y
40. Mendoza, J., Garrido, T., Cotoras, D., Arriagada, F.: Influence of agro-industrial wastes on the sorption, desorption, and distribution of copper in soils polluted by mine tailings. J Soils Sediments [Internet] **20**(3), 1571–1583 (2020 Mar 1) [cited 2023 Feb 21]. Available from: https://doi.org/10.1007/s11368-019-02507-z
41. Meza-Ramirez, V., Espinosa-Ortiz, X., Ramirez-Verdugo, P., et al.: Pb-Contaminated Soil from Quintero-Windows, Chile: Remediation Using Sarco cornianeei. Scientific World Journal [Internet] (2021) [cited 2023 Feb 22]. Available from: https://pubmed.ncbi.nlm.nih.gov/33679258/
42. Wong Arguelles, C., Carranza Alvarez, C.; Alonso-Castro, A.: Phytoremediation in situ in Mexico: A Review [Internet]. Magazine Fitotec, 1–10 (2021) [cited 2023 Feb 22]. Available from: https://revfitotecnia.mx/index.php/RFM/article/view/859/816
43. Casteblanco, J.A.: Techniques of remediation of heavy metals with potential application in the cultivation of cocoa. The Farm [Internet] **27**(1), 21–35 (2018) [cited 2023 Feb 22]. Available from: http://scielo.senescyt.gob.ec/scielo.php?script=sci_arttext&pid=S139085962018000100021
44. Nieves, Parra, N., Villanueva, S., Henriquez, M.: Tech note: bioremediation, enemy of cadmium / technical Note: bioremediation, the enemy of cadmium (2019) [cited 2023 Feb 21]; Available from: https://redalyc.org/journal/707/70758484010/html/
45. War Yepes, L.M.: Advances in the remediation of biological mercury: Fungi macroscopic as potential agents for bioremediation Advances in the biological remediation of mercury: Macroscopic fungi as potential bioremediation agents [Internet] (2019) [cited 2023 Feb 22]. Available from: http://repositorio.unal.edu.co/bitstream/handle/unal/77625/1128390092.2019.pdf?sequence=4
46. Mendarte-Alquisira, C., Alarcón, A., Ferrera-Cerrato, R.: Phytoremediation: an Alternative biotechnology to retrieve soil contaminated with DDT. A review. TIP Rev Spec Sciences Chemical-Biological [Internet]. **24**(Ddd), 1–15 (2021) [cited 2023 Feb 22]. Available from: https://redalyc.org/journal/432/43271644020/html/

47. Guzman Morales, A.R., Oriol Vázquez, P., Cross The Peace, Or., et al. Fitotecnología for the recovery of agro-ecosystems contaminated with heavy metals from industrial waste. Cent Investig Agropecu [Internet] **48**(3), 43–52 (2021). Available from: http://ref.scielo.org/h5fxp5
48. Brutti, L., Beltrán, M., Garcia de Salamone, Ines.: Bioremediation of Natural Resources [Internet]. INTA. Buenos Aires: Institute of Soils, Research Center of Natural Re-sources 1–45 (2018) [cited 2023 Feb 22]. Available from: https://ri.conicet.gov.ar/bitstream/handle/11336/135071/CONICET_Digital_Nro.47a8f79f-2c62-4398-9cc318e858fdc03e_H.pdf?sequence=11
49. Benavidez-Matias, E.: Selection of consortia native (plant-microorganisms that facilitate the rizorremediación of soils contaminated with heavy metals [Internet]. University of Buenos Aires (2021) [cited 2023 Feb 22]. Available from: https://bibliotecadigital.exactas.uba.ar/collection/tesis/document/tesis_n6832_Benavidez
50. Coyago, E., Bonilla, S.: Absorption of lead from soils highly contaminated species vegetative used for animal consumption, and human. The Farm [Internet] **23**(1), 35–46 (2016) Jun 30 [cited 2023 Feb 22]. Available from: http://revistas.ups.edu.ec/index.php/granja/article/view/23.2016.04
51. Chacon, C., Elizabeth, M., Casanca, M., Jennifer, R., Ordóñez-Sánchez, M., Alberto L.: Systematic Review: Bioremediation of soils contaminated by heavy metals (Pb, Cd, As, Ag, Cu, Ni, Zn) using native plants in areas too frequently **95** (2021) [cited 2023 Feb 21]. Available from: https://repositorio.ucv.edu.pe/handle/20.500.12692/103046
52. Cruzado-Tafur, E., Bierla, K., Torró Lisard, S.J.: Accumulation of As, Ag, Cd, Cu, Pb, and Zn by Native Plants Growing in Soils Contaminated by Mining Environmental Liabilities in the Peruvian Andes. Plants 2021 [Internet]. **10**(2), 241 (2021 Jan 27) [cited 2023 Feb 21]. Available from: https://www.mdpi.com/2223-7747/10/2/241/htm
53. Jaime Alberto Leal Afanador Rector, Yunda Perlaza Leonardo, Evemeleth Sánchez Torres, Vice president of Services Aspirants Leonardo, Alumni Edgar Guillermo Rodriguez Diaz, Vice-rector for International Relations Luigi Humberto López Guzmán Estudent Angel Osorio Julialba and Sebastian Chiriví Solomon Juan. Practice of bioremediation in soils and waters. Notes of Campus [Internet] **49** (2019 Apr 16) [cited 2023 Feb 21]. Available from: https://hemeroteca.unad.edu.co/index.php/notas/article/view/3451
54. Granollers Antoni.: Use of bacteria as an alternative of bioremediation in soils contaminated with Mercury (Hg), Zinc (Zn), Aluminum (Al). Copper (Cu) with the use of bacteria. Auhing Arcos, Javier; Guato Molina Jefferson, editor. Rev Ing UC [Internet] **80** (2022). Available from: https://repositorio.uteq.edu.ec/bitstream/43000/6757/1/7.pdf
55. Jara Medina julio enrique.: Development of a methodology for the recovery of soils highly contaminated with heavy metals using remediation physico-chemical and phytoremediation [Internet]. Andres Bello University. Andres Bello University; 2018. Available from: https://repositorio.unab.cl/xmlui/bitstream/handle/ria/7954/a122072_Jara_J_Desarrolloamethodology of recovery de_2018_Tesis.pdf?sequence=1&isAllowed=y
56. Wall Gonzales, D.: Characterization of plant species for fitorremediar soils contaminated with heavy metals in tailings mining [Internet]. Autonomous university of the State of Morelos; 2020. Available from: http://www.riaa.uaem.mx/xmlui/bitstream/handle/20.500.12055/1275/MUGDNL07T.pdf?sequence=1&isAllowed=y

Functionalization of Graphene Oxide with Poly(Amido Amine) and Its Application in Membrane Synthesis

Ahmed Toky Yasir[1,2], Abdelbaki Benamor[1,2(✉)], and Alaa H. Hawari[3]

[1] Gas Processing Center, College of Engineering, Qatar University, 2713 Doha, Qatar
benamor.abdelbaki@qu.edu.qa

[2] Department of Chemical Engineering, College of Engineering, Qatar University, 2713 Doha, Qatar

[3] Department of Civil and Architectural Engineering, College of Engineering, Qatar University, 2713 Doha, Qatar

Abstract. Graphene oxide (GO) suffers from excessive coagulation at high concentration which limits its use as membrane nanofiller. This issue can be tackled - among other options - by functionalizing GO with another nano-polymer. In this investigation, third generation of poly(amido amine) (PAMAM) was used to functionalize GO and the GO-PAMAM nanocomposite was incorporated in the polysulfone (Psf) membrane matrix. Synthesized GO- PAMAM was characterized using XRD, while the cross-sectional and surface morphology of the membranes were characterized using FE-SEM and AFM, respectively. The porosity, pore size and contact angle of the membrane were determined using appropriate techniques. The GO-PAMAM incorporated membrane exhibited a pure water permeation flux of 82 LMH/bar, which is 10 times higher than the flux exhibited by the Psf pristine membrane. Fouling tests on the new membrane showed a BSA rejection higher than 99% with an increased flux recovery rate of 10% higher compared to the one obtained using pristine Psf membrane. These preliminary results showed that blending GO-PAMAM nanocomposites in a polysulfone membrane matrix could be a good option to improve the performance of Psf ultrafiltration membranes.

Keywords: Graphene oxide · Poly(amido amine) · Wastewater

1 Introduction

70.9% of the earth is covered with water. However, 97% of the water is seawater and another 2.4% is glaciers and polar ice caps [1]. Other ware bodies, such as ponds and rivers, contain 0.6% of the water. Due to rapid urbanization and industrialization, the limited renewable source of water is under tremendous stress [2]. Wastewater reclamation and reuse can effectively reduce the stress on renewable fresh water source and aid in ensuring water security in the face of rapid urbanization and industrialization [3, 4]. One of the major wastewater reclamation technologies is membrane ultrafiltration process, which is favored since it effectively removes macromolecules and particulates

J. Baeyens et al. (Eds.): ESRE 2023, pp. 42–50, 2025.
https://doi.org/10.1007/978-981-96-4345-5_4

from wastewater under mild operating conditions [5, 6]. As ultrafiltration membrane technologies advance, it is anticipated that their application will grow and their operating costs will decrease even more. With prolonged operations, ultrafiltration membranes are becoming less cost-effective due to excessive irreversible membrane fouling, which also shortens their lifespan. One way to reduce membrane fouling is to improve its hydrophilicity. Hoseinpour et al. stated that hydrophilic nanomaterials with carboxyl groups (-COOH) and amine groups (-NH_2) are ideal for enhancing membrane characteristics [5]. Because of its spherical shape, high density of surface-active functional groups, biocompatibility, and structural homogeneity, poly (amido amine), or PAMAM, is one such nanoparticle that is advantageous for membrane modification [7]. On PAMAM's surface, there are a lot of amine groups, and as PAMAM generation increases, so does the number of amine groups on PAMAM's surface. Water diffusion is encouraged by this increase in the nanoparticle's active site availability and decreased structural compactness. Mansourpanah and companions. Developed PA active layer on PES membrane using Ag-GO-PAMAM nanocomposite that was synthesized [8]. Improved hydrophilicity, a 9% decrease in irreversible fouling, and an 11.9 percent increase in permeate flux were all observed in the modified membrane. PAMAM are soluble in water and must be anchored by another nanoparticle to prevent PAMAM leaching, despite the fact that PAMAM have significantly improved membrane flux and fouling [8, 9]. In order to prevent PAMAM from leaching from the membrane, graphene oxide (GO), one of the most renowned nanoparticles for membrane fabrication, forms covalent bonds with PAMAM [10]. Because GO has excellent hydrophilicity, high dispersibility in organic solvents, and high chemical stability, it is the perfect material for membrane modification [11].

Thus, in this study, GO was functionalized with PAMAM. The functionalized GO-PAMAM nanocomposite was used to synthesize hydrophilic polysulfone membranes. The membrane was analyzed in terms of pure water permeation flux and flux recovery rate.

2 Methodology

2.1 Materials

Sigma Aldrich provided the extra pure fine graphite powder, sodium nitrate ($NaNO_3$), potassium permanganate ($KMNO_4$), sulfuric acid (H_2SO_4), polysulfone pellets (Mw 35,000), N-Methyl-2-pyrrolidone (NMP), dimethylformamide (DMF), and bovine serum albumin (BSA). Dendritech, Inc. provided the Generation 3 PAMAM which was used. Distilled water was used throughout the experiment, and all of the chemicals were reagent grade.

2.2 Synthesis of Graphene Oxide

The graphene oxide (GO) was synthesized following the modified Hummer's methods [12]. The first step involved adding 115 mL of sulfuric acid to a round-bottomed flask containing 2.5 g of NaNO3 and 5 g of graphite powder. Once the mixture's temperature reached 10 °C, it was stirred for 30 min in an ice bath. Two hours were spent gradually adding 15 g of KMNO4 at this temperature. Next, for an hour, the temperature was raised to 35 °C. The mixture's internal temperature was then kept below 100 °C as 230 mL of distilled water was gradually added to the round-bottomed flask. Next, 300 mL of distilled water was added to the solution and it was stirred for an hour. To finish the oxidation process, 10 ml of 30% H2O2 were added to it. The mixture turned brilliant yellow when the remaining KMNO4 was reduced, signifying that the oxidation process was complete. After using a centrifugation setup to clean the mixture with water and HCl, freeze drying was used to produce GO powders.

2.3 Synthesis of GO-PAMAM Nanocomposite

In a round-bottom flask containing 0 mg of GO distributed in 37 ml of DMF, 1 mg of generation 3 PAMAM was added dropwise to allow for the functionalization of GO. During a full day at 80 °C, the mixture was agitated and refluxed. This step was finished, and then five centrifugations using ethanol and water were performed on the warm solution. The final step involved drying the product for six hours at 80 °C. Through XRD analysis, the synthesized GO-PAMAM nanocomposite was described.

2.4 Synthesis of GO-PAMAM Modified Polysulfone Membrane

Phase inversion synthesis was used to create the polysulfone (Psf) membranes integrated into GO-PAMAM. GO-PAMAM multifunctional composite was dissolved in 20% NMP to prepare the casting solution. After adding the polysulfone, the remaining 80% of NMP was continuously stirred for three hours at 300 rpm and 80 °C. After combining the two NMP solutions, they were ultrasonically sonicated for 30 min to guarantee that the nanocomposites were properly dispersed. Next, a tiny quantity of the ultrasonicated solution was applied to a glass plate, and a Filmographe doctor blade (Braive Instrument, Germany) was used to distribute it evenly. The phase inversion process was completed by moving the plate to a bath of distilled water. After removing the membrane from the glass plate, it was rinsed for 30 min with distilled water. The composition of different components used for the membrane fabrication process can be found in Table 1.

Table 1. Constituents of the fabricated membrane.

	NMP	Psf	GO-PAMAM*
Psf	83%	17%	0%
GP-0.1	83%	17%	0.10%
GP-0.3	83%	17%	0.30%

* Based on Psf wt.%

2.5 Membrane Characterization

The cross section of the pristine and modified membranes was examined using FESEM (Gemini model SUPRA 55VP-ZEISS, Oberkochen, Germany). To expose the cross-sections, the membranes were solidified and fractured under liquid nitrogen. The surface roughness of the pristine and modified membranes were analyzied using AFM.

The porosity (ε) of the membrane was determined using the gravimetric method and calculated as:

$$\varepsilon = \frac{\omega_1 - \omega_2}{A \times l \times d_w} \tag{1}$$

Here, ω_1, ω_2, A, l and d_w corresponds to weight of the water saturated membrane (gm), weight of the dry membrane (gm), surface area of the membrane (cm^2), thickness of the membrane (cm) and density of water (1 gm/cm^3), respectively. The pore radius (r_m) of the membranes were calculated using the porosity and Guerout–Elford–Ferry equation:

$$\sqrt{r_m} = \frac{(2.9 - 1.75\varepsilon)8\eta lQ}{\varepsilon \times A \times \Delta P} \tag{2}$$

Here, η, Q and ΔP corresponds to viscosity of water $\left(8.9 \times 10^{-4}\, Pa.s\right)$, volume of the permeated pure water per unit time $\left(m^3/s\right)$ and the operating pressure (Pa), respectively.

The permeation tests were conducted using a stirred dead-end cell operated at trans-membrane pressure of 3 bar. Before all the filtration studies, the membranes were compacted for 30 min at 3 bar. While operating the dead-end cell, the pure water permeate flux (J, LMH) was calculated as:

$$J = \frac{\mathrm{w_t} - w_o}{A \times d_w \times \Delta t} \tag{3}$$

Here, $\mathrm{w_o}$, w_t and Δt corresponds to initial mass of the permeate water (gm), mass of the permeate water at time t (gm) and filtration time (sec). The anti-fouling study was conducted using a solution of 2000 ppm BSA as model foulant. Using the data obtained during the fouling tests, the total fouling ratio (R_t) and flux recovery ration (FRR) was calculated as:

$$R_t = \frac{J_{w0} - J_{wf}}{J_{w0}} \times 100\% \tag{4}$$

$$FRR = \frac{\mathrm{J_{w1}}}{J_{w0}} \times 100\% \tag{5}$$

Here, J_{w0}, J_{wf} and $\mathrm{J_{w1}}$ corresponds to initial flux, flux after 1 h of fouling and flux after cleaning the membrane after 1t fouling cycle.

3 Results and Discussion

The XRD result for the synthesized GO and GO-PAMAM nanocomposite can be seen in Fig. 1. The XRD of GO shows a characteristic peak at 10.89° which corresponds to 8.32Å distance between the GO layers which is formed due to the presence of OH functional group on the surface of the GO nanoparticle [13]. The XRD result of PAMAM functionalized GO shows a significantly reduced peak which confirms the functionalization of GO with PAMAM. This also indicates the loss of crystallinity of the GO nanoparticle.

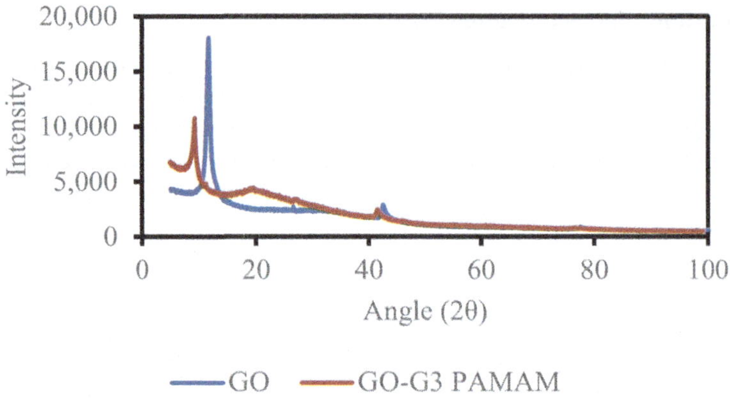

Fig. 1. XRD spectrum of GO and GO-PAMAM nanoparticle.

Figure 2 shows the porosity, pore radius and contact angle of the synthesized membranes. As seen from Fig. 2(a), the porosity of the pristine polysulfone (psf) membrane was 57%. Incorporating 0.1 wt% and 0.3 wt% of GO-PAMAM nanocomposite improved the porosity to 81% and 84%, respectively. Figure 2(b) shows that the pore radius of the pristine membrane was 13.5 nm, whereas the pore radius increased to 27.2 nm and 33.9 nm when 0.1 wt% and 0.3 wt% GO-PAMAM was incorporated, respectively. Finally, Fig. 2 (c) shows that the contact angle of the pristine membrane was 83° and reduced to 71° and 69° when 0.1 wt% and 0.3 wt% GO-PAMAM was incorporated. The presence of PAMAM in the casting solution reduces mass transfer resistance during phase inversion. This phenomenon is responsible for improved porosity and pore radius of the synthesized membraned. On the other hand, the presence of hydrophilic functional groups on the surface of PAMAM and GO is responsible for the improved hydrophilicity.

The cross-sectional FE-SEM images of the pristine and GO-PAMAM inorporated membranes can be seen in Fig. 3. As seen in Fig. 3, during the phase inversion process, finger-like pores has formed inside the memrbane. Moreover, the size of these channels became wider and deeper with increasing GO-PAMAM concentration from 0 to 0.3 wt%. This is because, during phase inversion, the GO-PAMAM nanocomposite are driven towards the non-solvent side from the solvent side owing to their hydrophilic nature.

The GO-PAMAM modified membranes and the pristine Psf membrane's topography and surface roughness are displayed in Fig. 4. In the figure the presence of crests on the membrane surface is indicated by the bright spots, while the presence of valleys and pores is indicated by the darker spots. Roughness values for the pristine membrane were 4.79 nm on average and 5.98 nm on the root mean square. When GO-PAMAM nanocomposite was added, it was discovered that both the average and root mean square roughness had increased.

Figure 5 shows the pure water permeation flux for the synthesized membranes. As seen from Fig. 4, the flux for the pristine psf membrane is 8.74 LMH/bar. Addition of 0.1 wt% GO-PAMAM nanocomposite improves the membrane flux by 527%, whereas

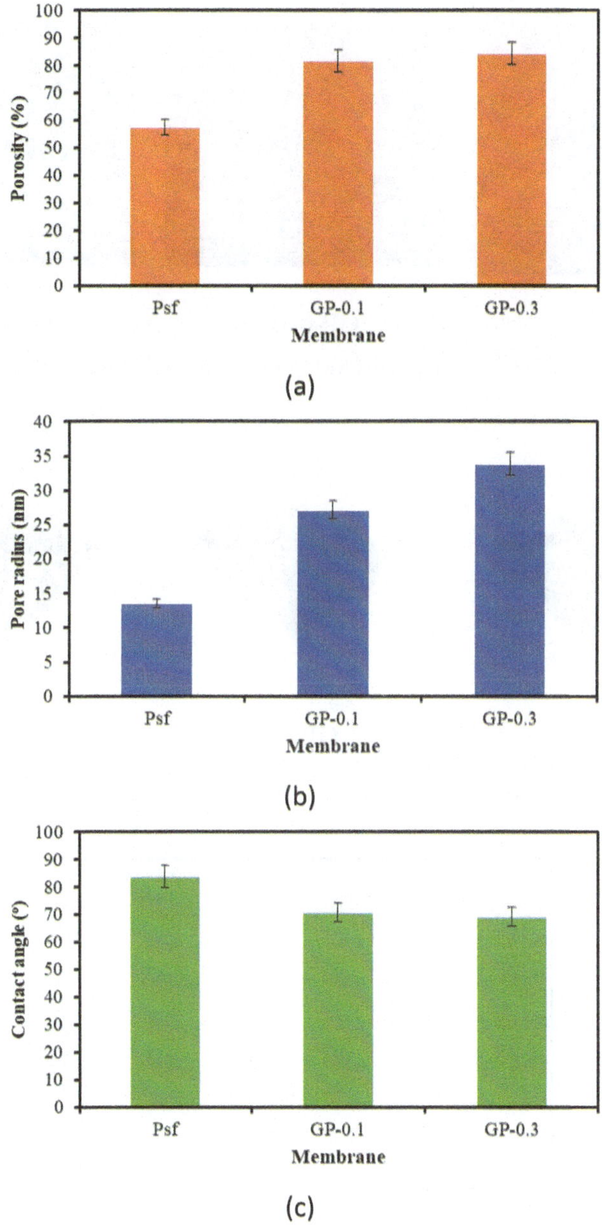

Fig. 2. (a) Porosity, (b) pore radius and (c) contact angle of the synthesized membranes.

addition of 0.3 wt% of GO-PAMAM nanocomposite improves the membrane flux by 943%. The improvement in porosity, pore size and hydrophilicity as discussed previously are the most significant factors that results in the improvement of membrane flux.

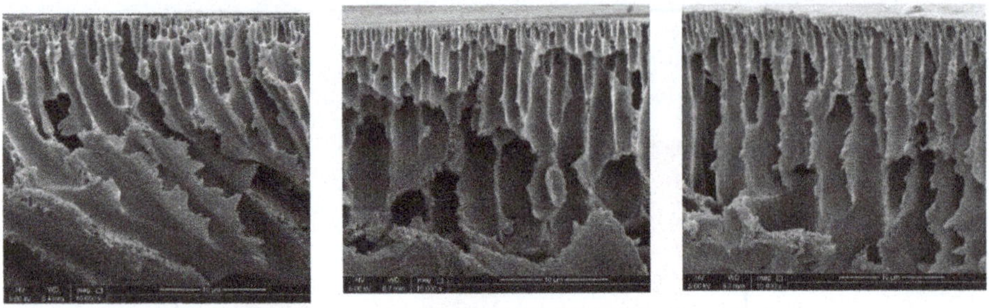

Psf GP-0.1 GP-0.3

Fig. 3. Cross sectional FE-SEM of the synthesized membranes.

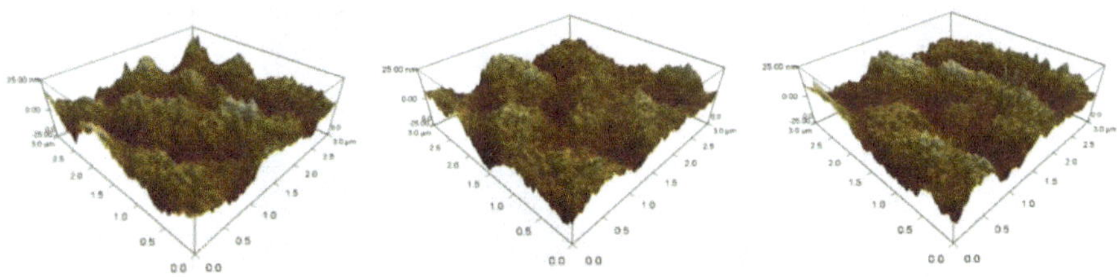

Psf GP-0.1 GP-0.3

Fig. 4. Surface roughness analysis of the pristine and modified membrane.

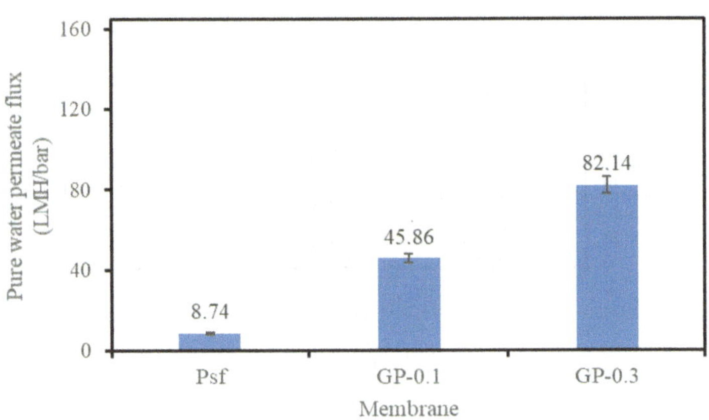

Fig. 5. Pure water permeates flux of the synthesized membranes.

Figure 6 shows the total fouling (R_t) and flux recovery rate (FRR) for the synthesized membranes. As seen from Fig. 5, the pristine membrane showed a total fouling and flux recovery rate of 98.9% and 17.39%. Addition of 0.1 wt% GO-PAMAM nanocomposite reduced fouling to 87.4% and increased flux recovery to 19.5%. Similarly, incorporating 0.3 wt% GO-PAMAM nanocomposite reduced total fouling to 87.0% and improved flux recovery rate to 28.6%. The improved hydrophilicity of the GO-PAMAM incorporated membranes are deemed responsible for the reduced total fouling and improved flux recovery rate.

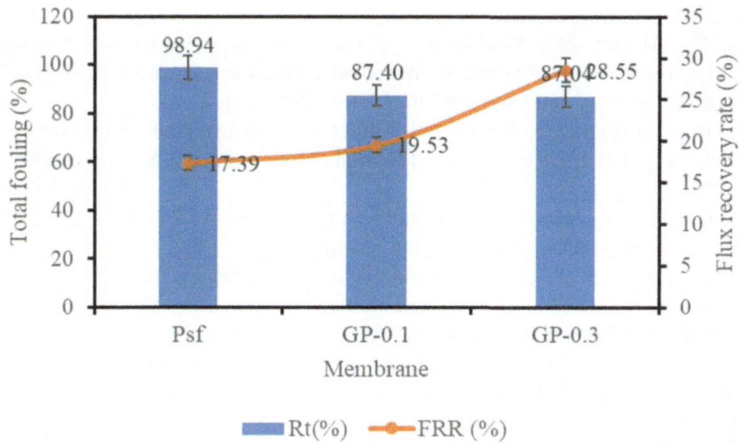

Fig. 6. Fouling properties of the synthesized membranes.

4 Conclusions

The proposed functionalization method was found to successfully functionalize GO with PAMAM. The functionalized GO-PAMAM nanocomposite was successful in improving porosity, pore radius and hydrophilicity of the membrane. Moreover, the GO-PAMAM incorporated membranes shows up to 8 times more membrane flux and significant reduction in membrane fouling parameters while maintaining higher than 99% BSA rejection rate.

Acknowledgement. This research is made possible by graduate sponsorship research award (GSRA7-1-0510-20046) from Qatar National Research Fund (QNRF). The statements made herein are solely the responsibility of the authors.

References

1. Aguilar, Y., Tadiosa, E., Tondo, J.: A Comparative Study on Wastewater Treatment Methods of Selected Multinational and Local Beverage Companies in the Philippines and Their Effects on the Environment. International Journal of Environmental Science and Development **5**, 570–574 (2014)

2. Ahmed, M., et al.: Potential Use Scenarios of Hybrid Biological Reactor for Petrochemical Industry Wastewater Treatment. International Journal of Environmental Science and Development **10**, 231–235 (2019)
3. Yasir, A.T., Eljack, F., Kazi, M.-K.: Synthesis of water capture technologies for gas fired power plants in Qatar. Chem. Eng. Res. Des. **154**, 171–181 (2020)
4. Hajrizi, F., Aliu, M.: Study of Physical-Chemical Characteristics of Discharge Wastewater in Lepenc Basin. International Journal of Environmental Science and Development **13**, 200–208 (2022)
5. Khajah, M., Ahmed, M., Al-Matouq, A.: Characterization of Reverse Osmosis Reject Wastewater Generated from Sulaibiya Wastewater Treatment Plant. International Journal of Environmental Science and Development **14**, 69–73 (2023)
6. Gamaralalage, D., Sawai, O., Nunoura, T.: Effectiveness of available rubber wastewater treatment facilities in Sri Lanka. International Journal of Environmental Science and Development **7**, 940–945 (2016)
7. Yasir, A.T., et al.: Poly (amido amine) Dendrimer Based Membranes for Wastewater Treatment - a Critical Review. Chemical Engineering Science, 118665 (2023)
8. Mansourpanah, Y., et al.: Silver-polyamidoamine/graphene oxide thin film nanofiltration membrane with improved antifouling and antibacterial properties for water purification and desalination. Desalination **511**, 115109 (2021)
9. Asempour, F., et al.: Chlorine-resistant TFN RO membranes containing modified poly(amidoamine) dendrimer-functionalized halloysite nanotubes. J. Membr. Sci. **623**, 119039 (2021)
10. Al-Gamal, A.Q., Saleh, T.A.: Design and manufacturing of a novel thin-film composite membrane based on polyamidoamine-grafted graphene nanosheets for water treatment. Journal of Water Process Engineering **47**, 102770 (2022)
11. Gosika, M., et al.: Covalent Functionalization of Graphene with PAMAM Dendrimer and Its Implications on Graphene's Dispersion and Cytotoxicity. ACS Applied Polymer Materials **2**(8), 3587–3600 (2020)
12. Mahmoudi, E., et al.: Novel nanohybrid polysulfone membrane embedded with silver nanoparticles on graphene oxide nanoplates. Chem. Eng. J. **277**, 1–10 (2015)
13. Mahmoudi, E., et al.: Distinguishing characteristics and usability of graphene oxide based on different sources of graphite feedstock. J. Colloid Interface Sci. **542**, 429–440 (2019)

Impact of Employing Weather Forecast Data as Input to the Estimation of Evapotranspiration by Deep Neural Network Models

Pedro J. Vaz[1,4], Gabriela Schütz[1,2], Carlos Guerrero[1,3], and Pedro J. S. Cardoso[1,4(✉)]

[1] Universidade Do Algarve, Faro, Portugal
pcardoso@ualg.pt
[2] CEOT—Centre for Electronics, Optoelectronics and Telecommunications, Faro, Portugal
[3] MED—Mediterranean Institute for Agriculture, Environment and Development, Évora, Portugal
[4] LARSyS—Laboratory of Robotics and Engineering Systems, Lisbon, Portugal

Abstract. Reference Evapotranspiration (ET_0) is a key parameter for designing smart irrigation scheduling, since it is related by a coefficient to the water needs of a crop. The United Nations Food and Agriculture Organization, proposed a standard method for ET_0 computation (FAO56PM), based on the parameterization of the Penman-Monteith equation, that is widely adopted in the literature. To compute ET_0 using the FAO56-PM method, four main weather parameters are needed: temperature, humidity, wind, and solar radiation (SR). One way to make daily ET_0 estimations for future days is to use freely available weather forecast services (WFSs), where many meteorological parameters are estimated up to the next 15 days. A problem with this method is that currently, SR is not provided as a free forecast parameter on most of those online services or, normally, such forecasts present a financial cost penalty. For this reason, several ET_0 estimation models using machine and deep learning were developed and presented in the literature, that use as input features a reduced set of carefully selected weather parameters, that are compatible with common freely available WFSs. However, most studies on this topic have only evaluated model performance using data from weather stations (WSs), without considering the effect of using weather forecast data. In this study, the performance of authors' previous models is evaluated when using weather forecast data from two online WFSs, in the following scenarios: (i) direct ET_0 estimation by an Artificial Neural Network (ANN) model, and (ii) estimate SR by (another) ANN model, and then use that estimation for ET_0 computation, using the FAO56-PM method. Employing data collected from two WFSs and a WS located in Vale do Lobo, Portugal, the latter approach achieved the best result, with a coefficient of determination (R^2) ranging between 0.893 and 0.667, when considering forecasts up to 15 days.

Keywords: Deep Learning · Public Garden · Smart irrigation

This work was supported by Project GSSIC—Green Spaces SMART Irrigation Control under Grant ALG-01-0247-FEDER-047030.

J. Baeyens et al. (Eds.): ESRE 2023, pp. 51–66, 2025.
https://doi.org/10.1007/978-981-96-4345-5_5

1 Introduction

One strategy to create a water-efficient irrigation system is to use soil humidity sensors to maintain humidity levels between the field capacity (FC) and the management allowable depletion (MAD), where MAD is a percentage of the soil's available water-holding capacity [1]. However, implementing these systems on public green spaces can be expensive due to the cost of components, installation, and management. Vandalism and theft can also be common issues, as these locations often lack comprehensive security or surveillance. Moreover, the soil's available water-holding capacity varies based on the soil type [2], requiring the use of laboratory analysis of samples to determine the humidity values corresponding to the field capacity and wilting point (WP) for a specific space.

Crop evapotranspiration (ET_c), also referred to as crop water use, is the water consumed by crops. The Food and Agriculture Organization of the United Nations (FAO) recommends using the FAO-56 Penman-Monteith (FAO56-PM) formula to calculate reference evapotranspiration (ET_0), which is related to crop evapotranspiration through a crop coefficient (K_c), i.e., $ET_c = K_c ET_0$ [3]. The FAO56-PM formula considers four main meteorological parameters: temperature, humidity, wind, and solar radiation (SR). The later parameter, SR, has been identified as the primary factor influencing ET_0 in several computational studies [4], but its measurement requires specialized sensors such as pyranometers, which are normally associated with expensive weather stations (WSs), and require specialized maintenance and calibration [5]. Furthermore, solar radiation forecast application programming interfaces (APIs) are also not widely available or can add significant costs to a system.

This paper is part of a framework for computing optimal irrigation schedules for crops, with a focus on green spaces. The framework uses computational models, based on machine learning [6] and deep learning [7], to estimate ET_0 using meteorological data from on-field weather stations and meteorological data (and forecasts) from internet weather forecast services (WFSs). This aims to optimize water and energy expenditure, improve crop health, reduce reaction time in addressing problems, enhance anomaly detection methods, and maintain the quality of green spaces. The framework, called Green Spaces SMART Irrigation Control (GSSIC), is being developed as an intelligent irrigation solution that is technologically differentiated from other platforms on the market.

Vaz et al. [6, 7] used data from a WS in Vale do Lobo, Portugal, to explore the use of machine learning and deep learning for ET_0 estimation. They concluded that the best results were obtained by using a deep learning regressor to estimate solar radiation (SR^{ANN}), using as input a limited set of meteorological features, and then using the SR estimate along with temperature, humidity, and wind speed as input to the FAO56-PM equation ($ET_0^{Hyb.}$). Using data from the Vale do Lobo WS, this approach achieved an R^2 value of 0.977, a root mean square error of 0.256 mm/day, a mean square error of 0.066 mm^2/day, a mean absolute error of 0.16 mm/day, and a mean absolute percentage error of 5.05%. In addition, good results were also obtained when directly estimating ET_0 using an Artificial Neural Network (ANN), ET_0^{ANN}, where an R^2 value of 0.959, a root mean square error of 0.342 mm/day, a mean square error of 0.117 mm^2/day, a mean absolute error of 0.25 mm/day, and a mean absolute percentage error of 7.54%

was achieved [7]. Figure 1 shows the workflow of the two ET_0 estimation approaches that were just detailed.

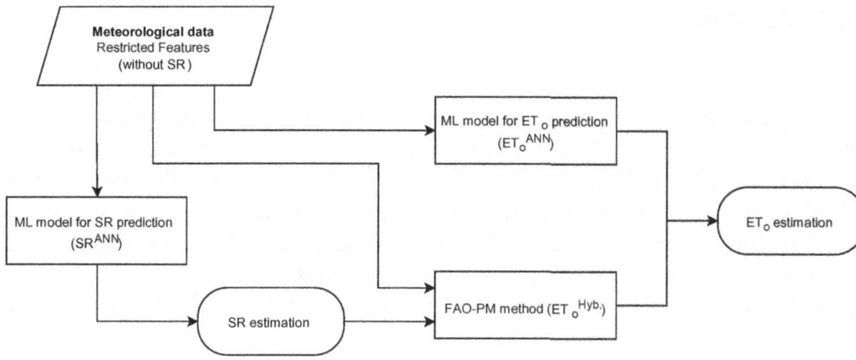

Fig. 1. Workflow of the two ET_0 estimation approaches: (i) ET_0 estimation model using a limited set of features (ET_0^{ANN}), and (ii) SR estimation model using a limited set of features (SR^{ANN}), and ET_0 computation based on using the estimated SR and a limited set of features as input to FAO56-PM formula ($ET_0^{Hyb.}$).

Importantly, the limited set of meteorological features that were used in both previously mentioned models are compatible with freely available WFSs, since SR is not required as an input feature. The current study builds on this work, by evaluating the performance of the deep learning models presented in [7], when using as input features historical weather forecast data gathered from two online WFSs, specifically Visual Crossing[1] (VC) and OpenWeatherMap[2] (OWM), for the Vale do Lobo Coordinates, allowing the assessment of the impact of using weather forecast data as input to the models.

The main contribution of this paper is the evaluation of the impact of using up to 15-day weather forecast data from online WFSs, as input to state-of-the-art ET_0 estimation models that are based on deep learning regressors [7]. As already described, two ET_0 estimation models are considered: one that directly estimates ET_0 from a set of restricted features (ET_0^{ANN}), and a hybrid model that uses a deep learning regressor to estimate SR and then applies the FAO56PM formula to compute ET_0 ($ET_0^{Hyb.}$). The key results are: (i) the $ET_0^{Hyb.}$ model performs better than the ET_0^{ANN}, (ii) both online WFSs that were considered (VC and OWM) give good and similar results for ET_0 estimation, and (iii) online weather forecasts can be used as inputs to the models and make ET_0 estimations of up to at least 11 days (d_{11}), at which point the R^2 metric falls below 0.7.

This paper is structured as follows. The next section presents a brief related work review of ET_0 estimation using observed and forecast data. The third section explores the used datasets, which includes data from a local WS and from two WFSs (VS and OWM), ending with the experimental setup details. Section 4 explores the impact on the

[1] https://www.visualcrossing.com.

[2] https://openweathermap.org.

performance of the proposed ET_0 and SR regressors when using up to 15-day forecast data. Finally, some conclusions and future work are presented in the last section.

2 Related Work

Reference evapotranspiration, ET_0, refers to the amount of water evaporated and transpired by a reference hypothetical grass surface with specific characteristics, including a uniform height of 0.12 m, a surface resistance of 70 sm^{-1}, and an albedo (reflection coefficient) of 0.23 [3]. Crop evapotranspiration, ET_c, represents the water needs of crops and is proportional to reference evapotranspiration through the crop coefficient, K_c [3]. Therefore, accurate prediction of ET_0 is crucial for smart irrigation scheduling as it determines the amount of water needed to be replenished during irrigation [8]. There are several factors that set ET_c apart from ET_0, including the crop cover density and total leaf area, the resistance of the foliage epidermis and soil surface to the flow of water vapor, the aerodynamic roughness of the crop canopy, and the reflectance of the crop and soil surface to short wave radiation [9]. With the value of K_c known, the ET_c can be estimated by $ET_c = K_c ET_0$.

It becomes evident that to accurately predict crop water requirements, accurate estimation of $ET_0$is crucial. Over the years, numerous deterministic methods have been devised to estimate reference evapotranspiration by utilizing either single or limited weather parameters and are generally classified into temperature-based, radiation-based, or combination-based methods [5]. Some of the more precise ones are the methods that combine temperature and radiation, such as the Penman [10], the modified Penman [11], and the FAO-56 Penman-Monteith (FAO56-PM) [3] formulas. Shahidian et al. [5] provide a comprehensive review of several of these methods, comparing their performance under different climate conditions. The authors found, that when applied to climates differing from the ones in which the methods were developed and tested, many of these methods showed poor performance and may necessitate the adjustment of empirical coefficients to suit local climates, which is not desirable.

Recently, machine and deep learning have been used as an alternative to estimate ET_0. Chia, Huang, and Koo [12] found that machine learning is a promising solution for ET_0estimation using common meteorological data. Granata [13] compared three evapotranspiration models and applied four machine learning algorithms (M5P Regression Tree, Bagging, Random Forest, and Support Vector Machine) to each model. The best results achieved a coefficient of determination of 0.987 and a mean absolute error of 0.14 mm/day. However, all models used net solar radiation as an input variable. Granata [14] also studied three recurrent neural network models for short-term evapotranspiration prediction, using Long Short-Term Memory (LSTM) and nonlinear autoregressive network with exogenous inputs algorithms. Ferreira et al. [8] compared six empirical reduced set equations with ANN and Support Vector Machine (SVM) models using data from 203 weather stations in Brazil for daily ET_0estimation. They found that ANN was the best-performing model when using data from up to four previous days as inputs, with a median R^2 value of around 0.80 for all stations.

Yang et al. [15] discuss the use of the Reduced-set Penman-Monteith model for short-term daily ET_0forecasting in eight weather stations in China. The model uses temperature data from 7-day public weather forecasts and wind speed data with four different types of wind speed inputs (default value, forecasted, long-term daily average, and annual average). For example, for one of the stations the mean absolute error for 7-day ET_0prediction ranged from 0.64 to 0.85 mm/day. Traore et al. [16] applied four ANN algorithms to predict ET_0in Dallas, Texas, by using restricted climate information messages retrieved from a public 15-day weather forecast source. The best ET_0performing model obtained a mean absolute error ranging from 0.767 to 0.996 mm/day, using as input features maximum and minimum temperature, and solar radiation. Luo et al. [17] tested four ANN models for ET_0 prediction using a 7-day temperature forecast from a public WFS. The obtained R^2 for 7-day ET_0prediction, using weather forecast data as input to the models, ranged from 0.46 to 0.61.

3 Datasets, Exploratory Analyses, and Experimental Setup

This section introduces the datasets used in this study with particular emphasis on the exploratory data analysis over the forecast data. It ends with a summary of the experimental setup.

3.1 Datasets

Data was collected from a WS located in Vale do Lobo, in south Portugal, as well as historical weather forecasts from the VC and OWM WFSs, using the Vale do Lobo WS coordinates, for years 2020 and 2022. We are not considering data from 2021 as one of the services had a gap of more than three months for our location over that year. The WS is equipped with sensors from Davis Instruments, which periodically measure and record various weather parameters at a daily resolution, including: temperature (minimum, maximum, and average), relative humidity (minimum, maximum, and average), solar radiation (maximum and average), wind speed (minimum, maximum, and average), atmospheric pressure (minimum, maximum, and average), rain intensity, and precipitation. The collected historical weather forecast data (from VC and OWM WFSs) includes 15-day historical forecasts, d0, d1,..., d15, where d0 represents today's forecast that was generated today (at midnight), d1 represents today's forecast that was generated yesterday, d2 represents today's forecast that was generated 2 days ago etc.

To evaluate and compare the performance of the models, the following statistical measures were used: mean absolute error (MAE), mean absolute percentage error (MAPE), mean square error (MSE), root-mean-square error (RMSE), and coefficient of determination (R^2) [18]. For reference, these metrics are defined as follows, given actual values y_t and estimated values $\hat{y}_t$, at instants $t = 1, 2, \ldots, n$, and the mean value of the actual samples $\overline{y}$: MAE $= \frac{1}{n}\sum_{t=1}^{n}\left|y_t - \hat{y}_t\right|$, MAPE $= \frac{1}{n}\sum_{t=1}^{n}\frac{\left|y_t-\hat{y}_t\right|}{|y_t|} \times 100\%$, MSE $= \frac{1}{n}\sum_{t=1}^{n}\left(y_t - \hat{y}_t\right)^2$, RMSE$= \sqrt{\text{MSE}}$ and $R^2 = 1-\sum_{t=1}^{n}\left(y_t - \hat{y}_t\right)^2/\sum_{t=1}^{n}(y_t - \overline{y})^2$. In this group of evaluation measures, MAE measures the average of the absolute differences between the original and predicted values, in the original unit, while MAPE measures the average absolute percentage error. MSE represents the average of the squared differences between the original and predicted values, which is a measure of the variance of the residuals in the squared unit of the original data. RMSE is the square root of MSE, which returns the residual variance to the original unit. Finally, R^2 stands for the proportion of the variance in the dependent variable that is explained by the linear regression model, and values should be close to 1 for a good model. On the other hand, values of MAE, MAPE, MSE, and RMSE should be close to 0 for a good model.

Depending on the type of problem being studied (estimating ET_0or SR), two targets were considered: (i) (daily) $ET_0$was calculated using the FAO56PM formula and input data from the WS, including temperature, humidity, wind speed, and solar radiation; (ii) the (daily) average solar radiation measured by the weather station was used as the target for solar radiation.

3.2 Exploratory Data Analysis of the Forecast Data

This section compares the historical weather forecast data, collected from Visual Crossing (VC) and OpenWeatherMap (OWM), from d0 up to d15, against the actual values measured by the Vale do Lobo WS, by means of the R^2 metric. Figure 2 presents the evolution of the R^2 metric from d0 up to d15, for maximum and minimum temperature [TempMax, TempMin], average humidity [HumidityAvg], and average wind [WindAvg]. These are relevant weather parameters since they are used as model input features to the models used in this study.

Independent of the feature, as predictable since a weather forecast that is issued today should be more accurate than one that was generated 15 days ago, R^2 has its higher values for d0 and gradually degrades as it approaches d15. Further, the best R^2 values are obtained for maximum and minimum temperature, ranging between 0.772 and 0.346, and between 0.855 and 0.408, respectively. It is also observable that, for these features, the initial R^2 values start to drop more sharply after d3. Furthermore, the R^2 curves have similar behavior for minimum and maximum temperature, dropping below 0.7 after d7 and d5, respectively. However, from all the forecast weather parameters, these are the more reliable for midterm forecast. For example, the R^2 values for average wind speed and average humidity are inferior to 0 after d6 and d8, respectively. With respect to average humidity, R^2 is very close for both weather service providers, but VC has a slightly higher R^2 value up to d3, where both curves match.

At d5, the R^2 value falls below 0.4, and at d8 falls below zero, which indicates that the average humidity forecasts should probably be used, at most, up to d3 or d4, for both services. Regarding average wind, starting at d0 with an R^2 of 0.540 and 0.354, for OWM and VC, respectively, both curves match at d4 with an R^2 below 0.2, indicating that from all the forecast weather parameters, average wind speed is the most unreliable, and should possibly be considered only up to d2. In general, for the Vale do Lobo WS coordinates, OWM has slightly higher R^2 values than VC (of course, this might have a different outcome for different locations). The impact of forecast values will be discussed further in the following sections.

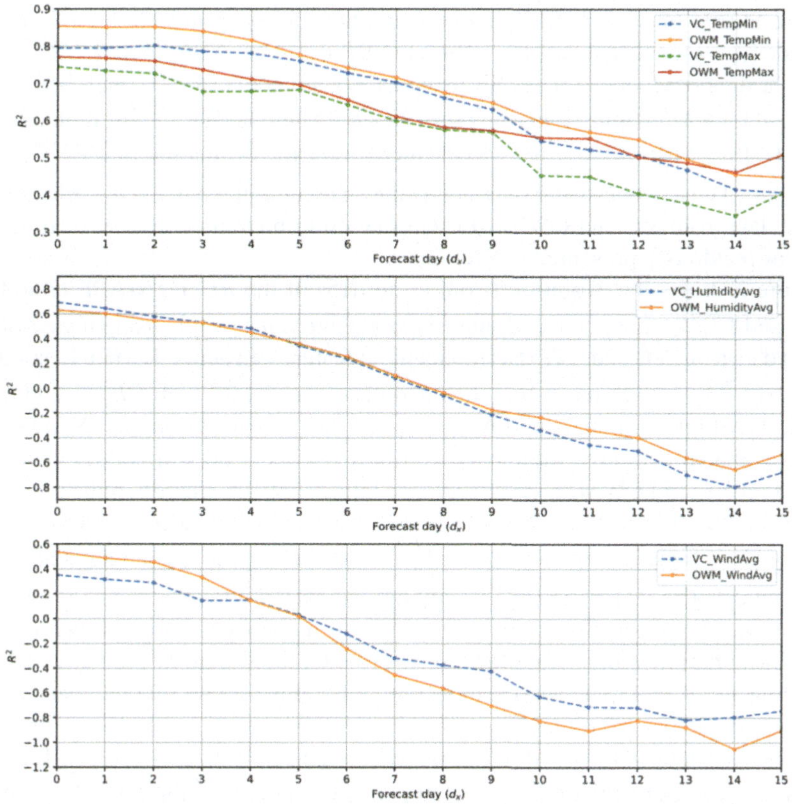

Fig. 2. Plots of the R^2 evolution for VC and OWM weather data in function of forecast day for maximum [TempMax] and minimum [TempMin] temperature (top), average humidity [HumidityAvg] (middle), and wind speed average [WindAvg] (bottom).

3.3 Experimental Setup

This study was performed in a computational environment using Python v3.9.7, Numpy v1.21.4 [19], Pandas v1.5.2 [20, 21], Tensor Flow v2.6.0 [22], Keras 2.6.0 [23], Scikit-learn v1.0.1 [24], and PyET v1.1.0 [25]. The Pandas library was employed for data analysis and manipulation, PyET to calculate ET_0using the FAO56PM method, Scikit-learn for data preprocessing and model metrics evaluation. All neural network models presented were created using Keras, which operates on top of Tensor Flow. Computation was performed on a 2020 MacBook Air equipped with an Apple M1 SoC chip and 16 GB of RAM, running MacOS Big Sur v11.6.4.

4 Impact of Using Forecast Data on $ET_0$and SR Regressors

In this section the impact of employing weather forecast data as feature inputs for $ET_0$and SR model inference is presented. To be clear, the models were not retrained using data from the WFSs, only inference was done. The models are thoroughly explained in [7], and were introduced in Sect. 1. More specifically, the following models for ET_0inference are used: (i) a model that directly estimates ET_0using an ANN model (ET_0^{ANN}), and (ii) a hybrid model that estimates SR using an ANN model (SR^{ANN}), and then uses that estimation for ET_0computation, using the FAO56-PM method ($ET_0^{Hyb.}$). Table 1 presents the R^2, RMSE, MSE, MAE, and MAPE metrics for the estimation of ET_0^{ANN}, SR^{ANN}, and $ET_0^{Hyb.}$. Those values were obtained using as input to the models historical weather forecast data from VC and OWM (for the Vale do Lobo WS coordinates), from d0 up to d15, as previously presented in Sect. 3.

Figure 3 shows the evolution (from d0 to d15) of the R^2 and MAPE plots for the ET_0^{ANN} and $ET_0^{Hyb.}$, obtained by doing ET_0model inference, while using meteorological forecast data from VC and OWM as model input features. In general, when looking at the R^2 plot (see Fig. 3, top), it can be concluded that up to d5, for both ET_0^{ANN} and $ET_0^{Hyb.}$, R^2 is slightly higher when using OWM forecast data. Then, starting at d5 and up to d9 the curves match and have similar R^2 values. R^2 Starts at d0 with values around 0.86 and 0.88 for VC and OWM, respectively, and falls below 0.8 after d6, indicating that for 6-day ET_0estimation, somehow trustful results can be obtained using forecast data from both providers. It is also noted that the $ET_0^{Hyb.}$ model always gives better results than the ET_0^{ANN} model. The R^2 for the $ET_0^{Hyb.}$ only falls below 0.7 after d11 and d13, for VC and OWM, respectively, suggesting that the proposed models can be employed for "mid" term ET_0estimation, which can play a major role on the design of predictive irrigation scheduling algorithms. The MAPE plot (Fig. 3, bottom), shows that the mean absolute percentage error for each of the ET_0models have similar values, when using either VC or OWM forecast data, with OWM only having a marginal advantage over VC. The MAPE plot also reinforces that the $ET_0^{Hyb.}$ model performs better than the ET_0^{ANN} model, with a difference between both of at least 2.5%.

Considering 25% as an admissible threshold for the MAPE, as before, a somehow trustful 6-day ET_0estimation can be obtained using forecast data from both providers. Lowering the threshold to 20% would suggest that the $ET_0^{Hyb.}$ predictions can be used for up to d4.

A set of violin plots of the MAE are presented on Fig. 4, where VC (blue), and OWM (orange) can be directly compared, for the ET_0^{ANN} and $ET_0^{Hyb.}$, as well as the SR^{ANN}models. When analyzing the plots, it can be seen that in general, and as expected, MAE increases as dx increases. It can also be seen that the models perform slightly better when using OWM data, but the difference is marginal.

These values are obviously worse than the ones reported by Vaz et al. [7] when using the observed (instead of forecast) Vale do Lobo WS test data. In that case, it was possible to attain an R^2 of 0.977 and 0.984, and a MAPE of 8.19 and 4.84%, for the ET_0^{ANN} and $ET_0^{Hyb.}$ models, respectively. To understand the impact of using weather forecast data on both ET_0estimation models, it is important to compare these values with the ones obtained with d0 weather forecast data. The averaged metrics (between VC, and OWM), give an R^2 value of 0.873 and 0.880, and a MAPE of 19.85 and 17.23%, for ET_0^{ANN} and $ET_0^{Hyb.}$, respectively. The conclusion is that, for the Vale do Lobo WS, a rule of thumb for the impact of using weather forecast data, is an approximate reduction of

Table 1. R^2, RMSE, MSE, MAE and MAPE metrics for ET_0^{ANN}, and $ET_0^{Hyb.}$, using historical weather forecast data collected from VC and OWM, as model input features. [Units for RMSE, MSE, and MAE are, respectively: *mm/day*, *mm²/day*, *mm/day* for *ETo*, and *W/m²/day*, *(W/m²)²/day*, *W/m²/day* for SR.]

		ET_o^{ANN}		ET_o^{Hyb}		SR^{ANN}	
Forecast	Day Metrics	VC	OWM	VC	OWM	VC	OWM
	R^2	0.859	0.886	0.866	0.893	0.849	0.858
	RMSE	0.704	0.633	0.688	0.614	33.331	32.251
d_0	MSE	0.496	0.401	0.473	0.378	1110.979	1040.145
	MAE	0.53	0.49	0.5	0.46	22.92	21.97
	MAPE	19.57	20.13	17.1	17.35	20.35	20.34
	R^2	0.846	0.881	0.854	0.887	0.847	0.858
	RMSE	0.736	0.648	0.718	0.632	33.435	32.238
d_1	MSE	0.542	0.42	0.515	0.399	1117.905	1039.272
	MAE	0.55	0.49	0.52	0.47	22.97	22.18
	MAPE	20.13	20.39	17.46	17.67	20.09	20.45
	R^2	0.837	0.868	0.847	0.878	0.836	0.843
	RMSE	0.76	0.683	0.735	0.655	34.588	33.924
d_2	MSE	0.577	0.466	0.54	0.429	1196.327	1150.869
	MAE	0.58	0.52	0.54	0.49	23.87	23.26
	MAPE	21.54	21.52	18.6	18.61	21.44	21.48
	R^2	0.821	0.861	0.835	0.871	0.831	0.842
	RMSE	0.796	0.701	0.765	0.675	35.019	34.012
d_3	MSE	0.634	0.492	0.585	0.455	1226.363	1156.814
	MAE	0.61	0.54	0.57	0.5	24.2	23.38
	MAPE	21.96	22.09	19.3	19.21	21.58	21.59
	R^2	0.816	0.841	0.828	0.853	0.822	0.828
	RMSE	0.803	0.748	0.778	0.72	35.989	35.453
d_4	MSE	0.645	0.56	0.605	0.519	1295.188	1256.935
	MAE	0.61	0.57	0.58	0.53	25.08	24.69
	MAPE	22.85	23.79	20.18	20.75	22.2	22.72
	R^2	0.827	0.835	0.846	0.849	0.817	0.823
	RMSE	0.779	0.764	0.735	0.731	36.527	35.983

(*continued*)

Table 1. *(continued)*

		ET_{o}^{ANN}		ET_{o}^{Hyb}		SR^{ANN}	
Forecast	Day Metrics	VC	OWM	VC	OWM	VC	OWM
d_5	MSE	0.607	0.584	0.541	0.534	1334.215	1294.809
	MAE	0.60	0.59	0.55	0.55	26.06	25.49
	MAPE	24.49	24.42	20.72	20.95	23.27	23.02
	R^2	0.803	0.803	0.817	0.816	0.801	0.804
	RMSE	0.833	0.834	0.803	0.806	38.147	37.867
d_6	MSE	0.694	0.695	0.645	0.65	1455.211	1433.914
	MAE	0.64	0.64	0.6	0.6	26.65	26.47
	MAPE	25.23	25.76	22.26	22.89	24.47	24.27
	R^2	0.76	0.766	0.782	0.786	0.775	0.788
	RMSE	0.919	0.909	0.874	0.869	40.407	39.361
d_7	MSE	0.845	0.826	0.765	0.756	1632.731	1549.282
	MAE	0.69	0.68	0.65	0.64	27.5	26.76
	MAPE	27.58	27.79	24.61	24.82	25.49	24.98
	R^2	0.745	0.739	0.762	0.757	0.762	0.767
	RMSE	0.946	0.959	0.915	0.927	41.615	41.318
d_8	MSE	0.896	0.921	0.838	0.859	1731.844	1707.188
	MAE	0.72	0.73	0.68	0.69	29.12	28.86
	MAPE	28.37	28.95	25.34	25.98	25.82	26.25
	R^2	0.719	0.71	0.734	0.727	0.732	0.748
d_9	RMSE	0.994	1.012	0.966	0.981	44.237	42.968
	MSE	0.9888	1.025	0.9333	0.963	1956.933	1846.234
	MAE	0.75	0.76	0.72	0.73	30.45	29.61
	MAPE	30.25	30.84	27.48	27.95	28.98	27.24
	R^2	0.682	0.703	0.71	0.728	0.709	0.727
	RMSE	1.055	1.024	1.006	0.98	45.961	44.679
$d10$	MSE	1.113	1.049	1.013	0.96	2112.418	1996.214

(continued)

Table 1. (*continued*)

		ET_o^{ANN}		ET_o^{Hyb}		SR^{ANN}	
Forecast	Day Metrics	VC	OWM	VC	OWM	VC	OWM
	MAE	0.81	0.79	0.76	0.74	31.2	30.1
	MAPE	32.14	32.79	28.97	29.35	29.19	28.53
	R^2	0.666	0.691	0.694	0.715	0.702	0.714
	RMSE	1.084	1.044	1.038	1.004	46.551	45.76
*d*11	MSE	1.175	1.091	1.078	1.008	2166.994	2093.95
	MAE	0.83	0.81	0.78	0.76	31.76	30.96
	MAPE	32.56	33.15	29.45	29.73	28.92	28.55
	R^2	0.672	0.683	0.695	0.704	0.718	0.725
	RMSE	1.075	1.058	1.037	1.023	45.367	44.865
*d*12	MSE	1.155	1.12	1.074	1.046	2058.132	2012.909
	MAE	0.83	0.83	0.79	0.78	32.04	31.45
	MAPE	32.81	34.27	29.9	30.87	28.99	28.67
	R^2	0.642	0.668	0.671	0.691	0.703	0.719
	RMSE	1.122	1.082	1.075	1.044	46.483	45.346
*d*13	MSE	1.26	1.171	1.157	1.089	2160.701	2056.296
	MAE	0.86	0.84	0.8	0.79	32.66	31.61
	MAPE	33.87	34.93	30.46	31.05	29.84	29.36
	R^2	0.624	0.644	0.662	0.677	0.7	0.713
	RMSE	1.15	1.121	1.09	1.068	46.587	45.818
*d*14	MSE	1.323	1.256	1.188	1.142	2170.357	2099.281
	MAE	0.9	0.87	0.84	0.82	33.14	32.25
	MAPE	34.79	35.66	31.56	32.03	29.87	29.73
	R^2	0.661	0.689	0.687	0.708	0.696	0.704
	RMSE	1.091	1.048	1.049	1.015	46.978	46.57
*d*15	MSE	1.189	1.099	1.1	1.03	2206.964	2168.786
	MAE	0.84	0.82	0.8	0.78	32.18	31.63
	MAPE	33.02	33.95	30.36	31.02	29.97	29.86

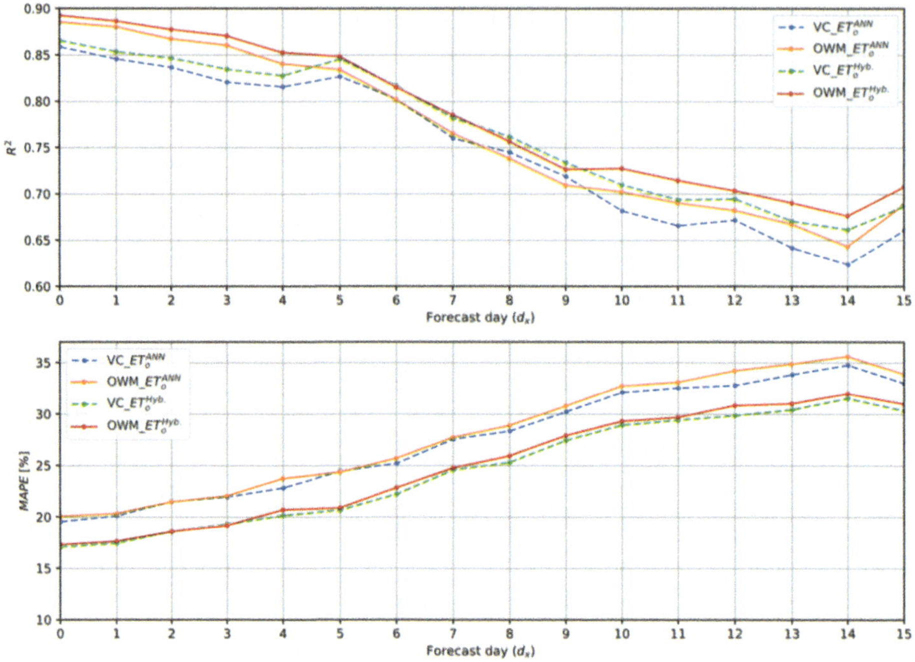

Fig. 3. R^2 (top), and MAPE (bottom) plots for ET_0^{ANN}, and $ET_0^{Hyb.}$, using historical weather forecast data collected from VC and OWM, as model input features.

0.1 on R^2, and an increase of 12% on the MAPE metric. Nevertheless, these values can improve the water efficiency of the green spaces by reducing or increasing the irrigation time according to the ET_0 predictions. Of course, this should also be integrated with precipitation, see Table 2 for an idea of the R^2 values associated to that forecast, being this out of the scope of this paper.

5 Conclusion and Future Work

Evapotranspiration is a key parameter for the design of predictive irrigation algorithms, that take into account future crop water needs, as well as precipitation and other meteorological parameter forecasts. In this study, and to the best knowledge of the authors, the impact and comparison of using two different online WFSs that provide 15-day forecast data, was for the first time assessed, when applied as input features to $ET_0$and SR estimation regressors based on deep learning. Despite the average performance of the metrics that are obtained when comparing the forecast meteorological data, with the data measured at the Vale do Lobo WS, especially regarding average humidity and wind, it was found that the ET_0regressors still gave good performance metrics, and it was shown that they can be used for midterm ET_0prediction, by using as input features data that is provided by common online and free WFSs. It was also concluded, that the ET_0estimation models had similar performance, when employing either VC or OWM

Table 2. R^2 metric comparing historical weather forecast data collected from VC and OWM weather forecast services against the measured values at the Vale do Lobo WS.

	TempMax		Templin		HumidityAvg		WindAvg		Preciptation	
	VC	OWM	VC	OWM	VC	OWM	VC	OWM	VC	OWM
d_0	0.746	0.772	0.796	0.855	0.695	0.632	0.354	0.54	–0.093	–0.039
d_1	0.735	0.769	0.796	0.852	0.646	0.605	0.319	0.49	–0.438	–0.111
d_2	0.728	0.761	0.803	0.853	0.58	0.547	0.292	0.457	–0.685	–0.545
d_3	0.679	0.737	0.787	0.841	0.531	0.53	0.147	0.334	–0.614	–0.407
d_4	0.68	0.712	0.782	0.817	0.485	0.451	0.15	0.148	–1.454	–0.907
d_5	0.684	0.697	0.761	0.778	0.344	0.36	0.031	0.022	–0.451	–0.367
d_6	0.644	0.656	0.729	0.743	0.239	0.257	–0.119	–0.242	–1.238	–1.106
d_7	0.600	0.611	0.704	0.717	0.082	0.103	–0.315	–0.455	–1.57	–1.388
d_8	0.576	0.582	0.662	0.676	–0.059	–0.037	–0.37	–0.562	–0.943	–0.856
d_9	0.569	0.573	0.631	0.649	–0.215	–0.174	–0.426	–0.703	–1.145	–1.006
$d10$	0.453	0.554	0.546	0.597	–0.337	–0.234	–0.631	–0.826	–0.868	–1.358
$d11$	0.45	0.552	0.522	0.569	–0.455	–0.337	–0.712	–0.904	–0.962	–1.299
$d12$	0.405	0.502	0.507	0.549	–0.504	–0.398	–0.718	–0.822	–1.268	–0.865
$d13$	0.379	0.487	0.468	0.496	–0.696	–0.56	–0.817	–0.877	–1.898	-2.273
$d14$	0.346	0.462	0.416	0.456	–0.793	–0.654	–0.793	–1.05	–1.031	–1.45
$d15$	0.406	0.51	0.408	0.449	–0.673	–0.529	–0.743	–0.902	–1.283	–1.306

weather forecast data and that the hybrid $ET_0^{Hyb.}$ model has better performance than the ET_0^{ANN} model, when using weather forecast data. This was expected and in line with previous findings by the authors.

Future work will include the integration of the precipitation forecast in a formula to compute the irrigation time, the generalization of these results to more WSs, as well as retrain the models using (d0) weather forecast data as features, instead of the WS measured features, to further try to improve ET_0 estimation model performance when using freely available online weather forecast services.

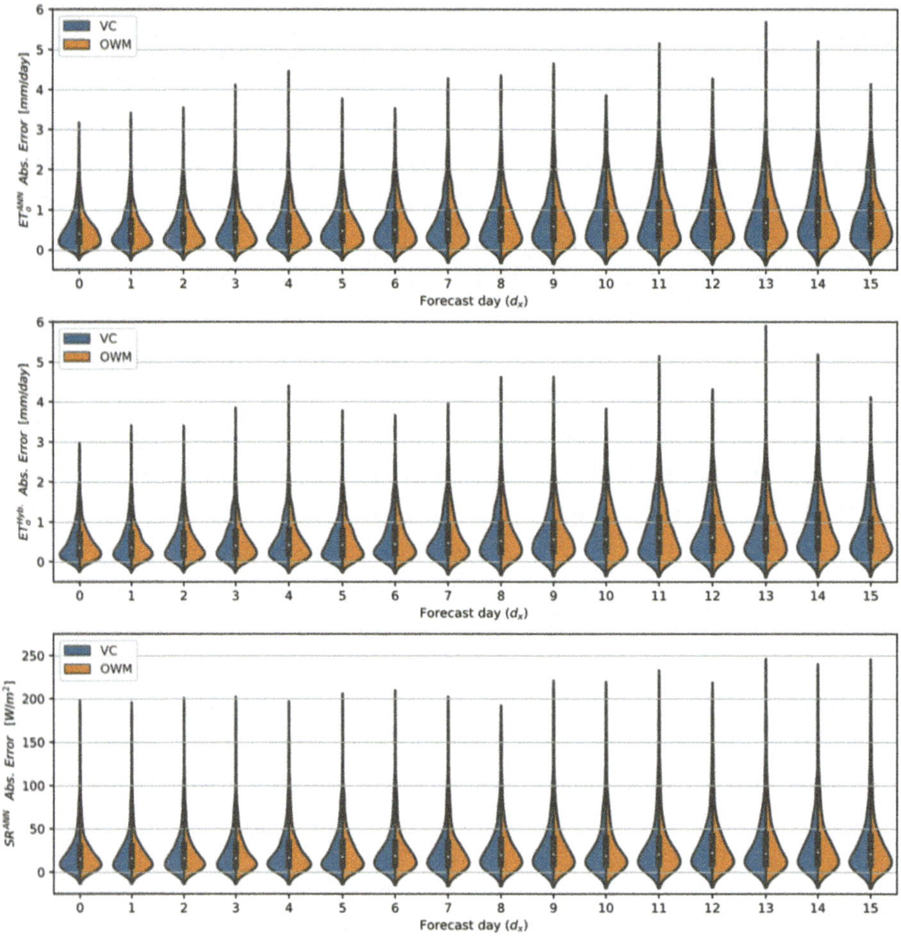

Fig. 4. Violin plots showing absolute error for ET_0^{ANN} (top), $ET_0^{Hyb.}$ (middle), and SR^{ANN} (bottom), in function of forecast day (d0, d1,..., d15), and forecast data collected from VC and OWM.

Acknowledgements. The authors thank the GSSIC Project's Companies, Visualforma - Tecnologias de Informação, S. A. and Itelmatis, Lda., and the Portuguese Foundation for Science and Technology (FCT) under Project UIDB/50009/2020—LARSyS, Project UIDB/00631/2020–CEOT BASE, Project UIDP/00631/2020—CEOT Programático, and Project UIDB/05183/2020–MED. We also wish to thank Visual Crossing for offering the data for this study.

References

1. Zhang, J., et al.: Sustainable irrigation based on co-regulation of soil water supply and atmospheric evaporative demand. Nat. Commun. **12**(1), 1–10 (2021)
2. Jong, R.D., Shields, J.: Available water-holding capacity maps of Alberta, Saskatchewan and Manitoba. Can. J. Soil Sci. **68**(1), 157–163 (1988)

3. Allen, R.G., et al.: Crop evapotranspiration guidelines for computing crop water requirements-FAO irrigation and drainage paper 56. FAO, Rome **300**(9), D05109 (1998)
4. Valipour, M.: Study of different climatic conditions to assess the role of solar radiation in reference crop evapotranspiration equations. Arch. Agron. Soil Sci. **61**(5), 679–694 (Jul 2014). https://doi.org/10.1080/03650340.2014.941823
5. Shahidian, S., Serralheiro, R., Serrano, J., Teixeira, J., Haie, N., Santos, F.: Hargreaves and other reduced-set methods for calculating evapotranspiration. IntechOpen (2012)
6. Vaz, P.J., Schütz, G., Guerrero, C., Cardoso, P.J.S.: A study on the prediction of evapotranspiration using freely available meteorological data. In: Proceedings of the International Conference on Computational Science. London (accepted) (21–23 June 2022)
7. Vaz, P.J., Schütz, G., Guerrero, C., Cardoso, P.J.S.: Hybrid neural network based models for evapotranspiration prediction over limited weather parameters. IEEE Access **11**, 963–976 (2023). https://doi.org/10.1109/ACCESS.2022.3233301
8. Ferreira, L.B., da Cunha, F.F., de Oliveira, R.A., Filho, E.I.F.: Estimation of reference evapotranspiration in Brazil with limited meteorological data using ANN and SVM: A new approach. J. Hydrol. **572**, 556–570 (2019). https://doi.org/10.1016/j.jhydrol.2019.03.028
9. Allen, R.: Crop coefficients. Encyclop. Water Sci. (2003)
10. Penman, H.L.: Natural evaporation from open water, bare soil and grass. Proc. Royal Soc. London. Ser. A. Math. Phys. Sci. **193**(1032), 120–145 (1948)
11. Doorenbos, J.: Guidelines for predicting crop water requirements. FAO Irrig. Drain. Paper **24**, 1–179 (1977)
12. Chia, M.Y., Huang, Y.F., Koo, C.H.: Resolving data-hungry nature of machine learning reference evapotranspiration estimating models using inter-model ensembles with various data management schemes. Agric. Water Manag. **261**, 107343 (2022)
13. Granata, F.: Evapotranspiration evaluation models based on machine learning algorithms. Agric. Water Manag. **217**, 303–315 (2019)
14. Granata, F., Di Nunno, F.: Forecasting evapotranspiration in different climates using ensembles of recurrent neural networks. Agric. Water Manage. **255**, 107040 (2021). https://doi.org/10.1016/j.agwat.2021.107040
15. Yang, Y., et al.: Short-term forecasting of daily reference evapotranspiration using the reduced-set penman-monteith model and public weather forecasts. Agric. Water Manag. **211**, 70–80 (2019)
16. Traore, S., Luo, Y., Fipps, G.: Deployment of artificial neural network for shortterm forecasting of evapotranspiration using public weather forecast restricted messages. Agric. Water Manag. **163**, 363–379 (2016)
17. Luo, Y., et al.: Medium range daily reference evapotranspiration forecasting by using ANN and public weather forecasts. Water Resour. Manage **29**, 3863–3876 (2015)
18. Chicco, D., Warrens, M.J., Jurman, G.: The coefficient of determination R-squared is more informative than SMAPE, MAE, MAPE, MSE and RMSE in regression analysis evaluation. PeerJ Comput. Sci. **7**, e623 (2021)
19. Harris, C.R., et al.: Array programming with NumPy. Nature **585**(7825), 357–362 (2020). https://doi.org/10.1038/s41586-020-2649-2
20. McKinney, W.: Data structures for statistical computing in python. In: Stéfan van der Walt, Jarrod Millman (eds.) Proceedings of the 9th Python in Science Conference. pp. 56–61 (2010). https://doi.org/10.25080/Majora-92bf1922-00a
21. Pandas Development Team, T.: pandas-dev/pandas: Pandas (Feb 2020). https://doi.org/10.5281/zenodo.3509134
22. Abadi, M., et al.: TensorFlow: Large-scale machine learning on heterogeneous systems (2015), software available from https://www.tensorflow.org/
23. Chollet, F.: Keras. https://github.com/fchollet/keras (2015)

24. Pedregosa, F., et al.: Scikit-learn: Machine learning in Python. J. Mach. Learn. Res. **12**, 2825–2830 (2011)
25. Vremec, M., Collenteur, R., X.Z.: PyEt – Open source python package for calculating reference and potential evaporation (v1.0.1). Zenodo (2021)

Chloride and Sulfate Resistance of the Alternative Raw Materials-Based Geopolymer

Sirirat Tubsungnoen Rattanachan(✉), Rawee Dangwiriyakul, Oranich Thongsri, Paritat Thaitalay, and Sawitri Srisuwan

School of Ceramic Engineering, Institute of Engineering, Suranaree University of Technology, 111 University, Avenue, Muang District, Nakhon Ratchasima 30000, Thailand
sirirat.b@g.sut.ac.th

Abstract. The service life of concrete used in acid or salt environments affects the salt resistance of cement and concrete. This project aims to study the change of chemical structure and microstructure of the geopolymer composition under soaking in high concentrations of chloride and sulfate at 50,000 mg/L as the simulated salt solution in this project. Geopolymer consisted of 10%wt ordinary Portland cement (OPC type 1) with the other alternative raw materials which were pozzolanic materials and wastes (fly ash and silica fume). The ion penetration into the geopolymer samples after soaking in the simulation salt solution was determined to understand the salt resistance of geopolymer as compared to ordinary Portland cement. Moreover, this geopolymer showed that the chloride and sulfate penetration into the samples presented significantly lower on the inside of the sample than that of the sample surface as compared to the Portland ordinary cement. In addition, the phases change after the setting of geopolymer in high concentrations of chloride and sulfate solution were studied for understanding the mechanism of salt resistance of geopolymer and for developing the concrete composition for salt resistance in the future.

Keywords: Geopolymer · Salt resistance · Cement Corrosion

1 Introduction

Those areas can harm the environment and buildings which degrade the mechanical properties and decrease life expectancy of structural engineering. Building materials with high salt resistance are interesting to develop and improve the quality of ordinary building materials. The global warming with issue of CO_2 emissions by cement production trend to the development of geopolymer. Geopolymer is three dimensional amorphous to semi-crystalline of aluminosilicate formed from the raw materials rich in silica and aluminium [1]. Geopolymer appear as a prospective construction material for the future due to their environmentally beneficial technology [2]. The replacement ratios from the cement content were 10, 20 and 30% respectively, for each recycling of rice husk ash (RHA) and fly ash (FA) from power plants as reactivity materials for producing

J. Baeyens et al. (Eds.): ESRE 2023, pp. 67–75, 2025.
https://doi.org/10.1007/978-981-96-4345-5_6

sustainable concrete. The water permeability loss ratios increased as the cementitious content decreased [3]. This result achieved the main goal by using high replacement ratios of RHA and FA for improving the durability of concrete. Compressive strength increases with the increase in the percentage of Fly ash and Rice Husk Ash up to replacement (22.5% FA and 7.5% RHA) of Cement in Concrete for different mix proportions [4]. The previous study was expected to provide a practical solution of utilizing FA and RHA solid waste in construction activities [5]. Results showed that the geopolymer concretes exhibited better corrosion resistance than OPC concrete. The higher the amount of Na_2SiO_3 and higher the concentration of NaOH solutions, the better the corrosion resistance of geopolymer concrete was [6]. The test results confirmed that the mechanical properties of the geopolymer mixtures tested are competitive with those of OPC concrete and provide a more durable alternative to the OPC concrete in a seawater environment [7].

In this study, it aims to study chloride and sulphate resistance of the geopolymer by the other alternative raw materials such as fly ash (FA), silica fume. Chemical and phase analysis were provided for evaluating the reaction and the change of chemical composition during soaking in the high concentrations of chloride and sulphate solution as compared to the ordinary Portland cement. It will bring the understanding of the corrosion mechanism and develop the new cement compositions.

2 Materials and Methods

Ordinary Portland cement under the brand name "Siam Cement Group", was utilized as a cement or binding material. Portland cement type 1 (OPC1) was the general cement. Fly ash (FA) was collected from Mae Moh coal power plant, Lampang province in Thailand. The Mae Moh coal power plant generated about 20,000,000 Tons of fly ash annually, which are dumped in a landfill. The silica fume was purchased from Elkem. Particle size distribution of all raw materials were characterized by using laser scattering particle size distribution analyzer (LA-950V2/Horiba) with a step time of 0.02°/min and a scan range of 10–70°. Chemical analysis of all raw materials was evaluated by X-ray fluorescence spectrometer (D2-phaser/ Bruker).

The cement composition was composed of ordinary Portland cement with the alternative raw materials for replacement of the ordinary Portland cement content such as fly ash and silica fume for 10%wt. The liquid solution consisted of NaOH (10M): Na_2SiO_3 in ratio of 24: 16 (wt./wt.) solution. The liquid to powder ratio was 0.4 in the range. Geopolymer compositions were studied in this work as summarized in Table 1.

Geopolymer samples were prepared from the mixture of OCP and the alternative raw materials with the alkali solution. The cubic shape specimens with 50 mm × 50 mm in dimension were prepared for tests. Then, the samples were kept in mold for 1 day before demolded and wrapped by the plastic bag for curing in the hot oven at 60 °C for 7days. After 7 days of curing, the samples were soaked in different conditions (normal water and the simulated salt) for 14 days. The simulated salt was prepared from the solution of sodium chloride (NaCl) and magnesium sulphate ($MgSO_4$) for concentration of chloride and sulphate of 50,000 mg/L. After soaking in the different conditions for 14 days, the samples were dried in air for 1 day.

Table 1. Composition of geopolymer samples

Sample code no	Powder (g)			Liquid (g)		
	OPC1	Fly ash	Silica fume	Water	NaOH (10M):Na_2SiO_3 (wt./wt.)	L/P ratio
T1	100			40	–	0.4
FA	10	90			40	0.4
FASiO2	10	85	5		40	0.4

For further analysis, the outer and inner of all samples were characterized for chemical analysis by using X-ray diffractometer (XRD) technique to evaluate the effect of the alternative raw materials and geopolymer reactions on the different soaking conditions. An X-ray diffractometer (XRD, D2-phaser/Bruker) was used to analyze the components and crystalline phase variation of geopolymer with the Cu Kα radiation ($\lambda = 1.5406$ Å). The data were analyzed in the range of 2θ from 10 to 70, and in increments of 0.02 with the counting time per step of 0.5. The morphologies of geopolymer soaked in the different conditions were observed using Scanning electron microscopy with energy dispersive X-ray (SEM/EDX, JSM-6010LV/JEOL) at the accelerating voltage of 15 kV.

3 Results and Discussion

Chemical compositions of all raw materials in this study were characterized by X-ray fluorescence spectroscopy (XRF) as summarized in Table 2. OPC type 1 presented high CaO and SiO_2 which were mainly acted as binder pozzolanic reaction. Fly ash provided high Al_2O_3 and SiO_2 contents. Silica fume presented mainly high amorphous silica of 93.74%. Fly ash and silica fume have average particle size of 4032 and 35.81 micron, respectively.

The chemical phase analysis of geopolymer samples was studied for understanding the hardening reactions as well as chloride and sulfate resistance as compared to the ordinary cement T1. The inside and outside of the sample soaked in the different solutions were characterized by SEM-EDX technique for microstructural examination and chemical analysis. Phases analysis of the ordinary cement T1 presented in XRD patterns as shown in Fig. 1.

The main phases of T1 soaked in water for 14 days were $CaCO_3$ and $Ca(OH)_2$, especially inside of the samples showed much $Ca(OH)_2$ (Fig. 1(d)). It found that the phase difference of T1 soaked in the salt solution (as shown in Fig. 1(a)) which occurred $MgOH_2$ phase on the surface sample and much of $Ca(OH)_2$ as compared to those samples soaked in water.

Phase analysis of FA geopolymer inside and outside samples soaked in the different solutions resulted not significantly different among them as shown in Fig. 2(a)-(d). While phase analysis of FASiO2 geopolymers inside and outside samples soaked in the different solutions also showed not significantly difference (Fig. 3) but it can be noticed that less $Mg(OH)_2$ was found on the outside FASiO2 soaked in the salt solution for 14 days (Fig. 3(a)).

Table 2. Chemical analysis of raw materials for geopolymers

Elements (%)	Raw materials		
	OPC I	Fly ash (FA)	Silica fume (SiO_2)
SiO_2	22.3	57.88	93.74
Al_2O_3	2.87	29.36	–
Fe_2O_3	4.03	5.52	4.04
CaO	60.36	3.19	0.88
Na_2O	2.81	–	–
K_2O	–	2.84	0.67
MgO	2.61	–	–
SO_3	3.03	–	–
TiO_2	0.25	–	–
Cr_2O_3	0.04	–	–
MnO_2	0.13	–	–
CuO	0.05	–	–
ZnO	0.45	–	–
SrO	0.27	–	–
Yb_2O_3	0.44	–	–

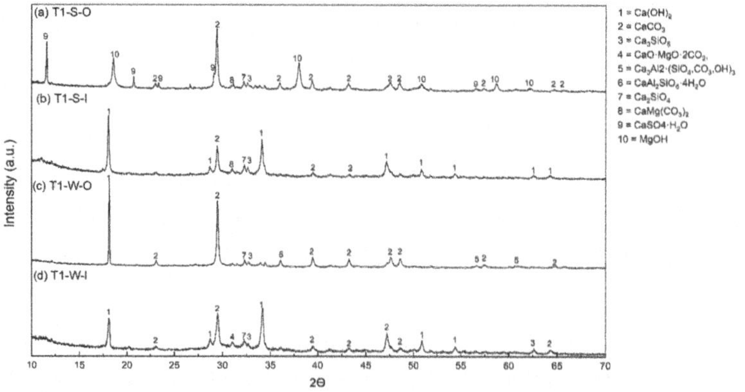

Fig. 1. XRD patterns of (a) surface, (b) inside of T1 sample soaked in the salt solution for 14 days and (c) surface and (d) inside of T1 sample soaked in water for 14 days.

Figures 4, 5 and 6 show the results of SEM-EDX micrographs of inside and outside samples for T1 soaked in the salt solution. After soaking in the salt solution, it found that Mg content mainly appeared on the surface. It should be noted that there were some contents of Na, Cl, and S on the outside of sample, and Mg content remarkably

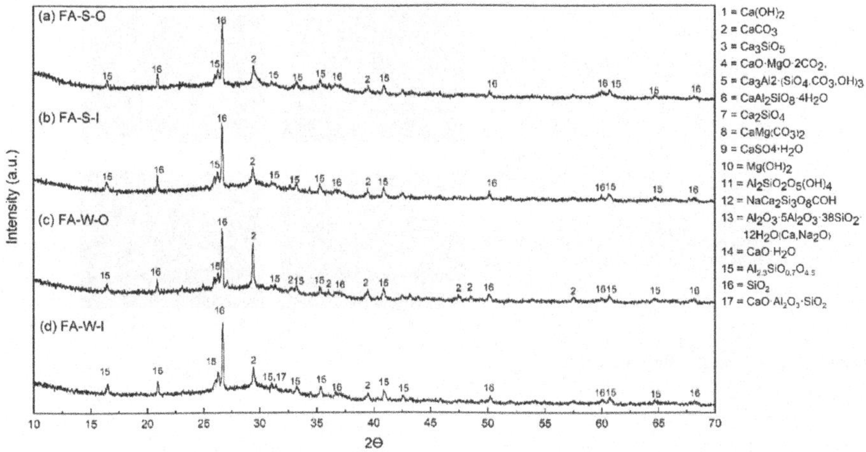

Fig. 2. XRD patterns of (a) surface, (b) inside of FA geopolymer soaked in the salt solution and (c) surface and (d) inside of FA geopolymer soaked in water for 14 days.

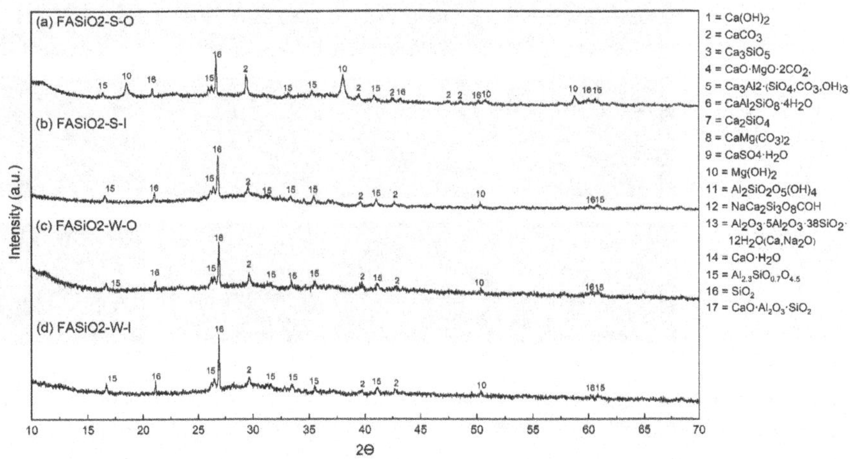

Fig. 3. XRD patterns of (a) surface, (b) inside of FASiO2 geopolymer soaked in the salt solution and (c) surface and (d) inside of FASiO2 geopolymer soaked in water for 14 days.

increased on outside samples soaked in the salt solution (Fig. 4(b)). Interestingly, SEM-EDX results for both of inside and outside samples for FA and FASiO2 geopolymers showed no notable change after soaking in the salt solution as shown in Figs. 5 and 6, respectively.

It should be noted that FASiO2 should be adjusted the liquid and powder ratio for obtaining high densified body and its potential used for geopolymer with good salt resistance.

Research has demonstrated that geopolymer concrete exhibits long-term compressive strength even in aggressive environments. The rate of chloride diffusion within

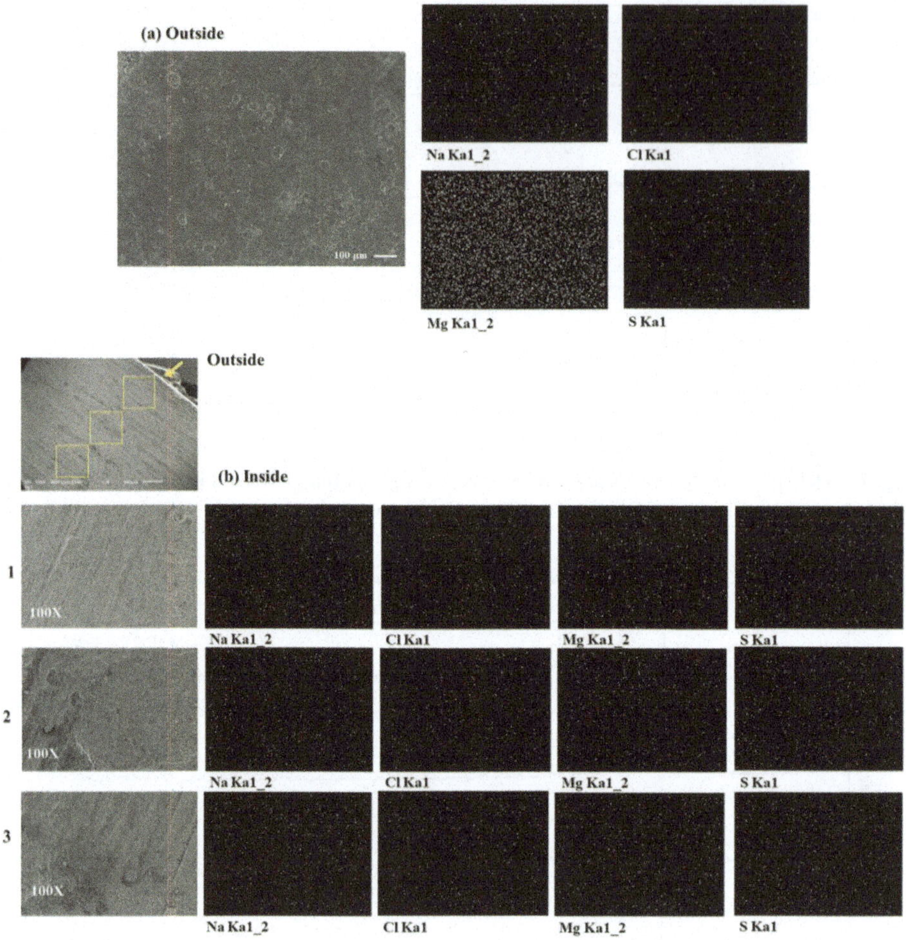

Fig. 4. SEM-EDX micrographs of (a) outside, (b) inside of T1 sample soaked in the salt solution for 14 days (1,2 and 3 were analyzed from outside to inside of sample, respectively).

geopolymer concrete hinges on the porosity of its matrix, which, in turn, is influenced by the reactivity of the precursor fly ash. This reactivity depends on many factors such as surface area, amorphous Al_2O_3 and SiO_2 content, and CaO content. Greater surface area and concentrations of amorphous Al_2O_3 and SiO_2 promote aluminosilicate gel formation, yielding a stronger and less porous geopolymer matrix, thereby mitigating chloride ingress [8]. During the geopolymerization process, silica and alumina in fly ash dissolve with NaOH to yield sodium-alumina-silicate, enhancing the strength of geopolymer concrete. Conversely, the compressive strength of Ordinary Portland Cement (OPC) diminishes due to the leaching of portlandite ($Ca(OH)_2$), which reacts with NaCl to form calcium chloride ($CaCl_2$). This compound further reacts with aluminate in the cement, forming Friedle salt ($3CaO.Al_2O_3.CaCl_2.10H_2O$), which fills pores and may lead to cracks and salt crystal accumulation [9]. Similar effects were observed in the

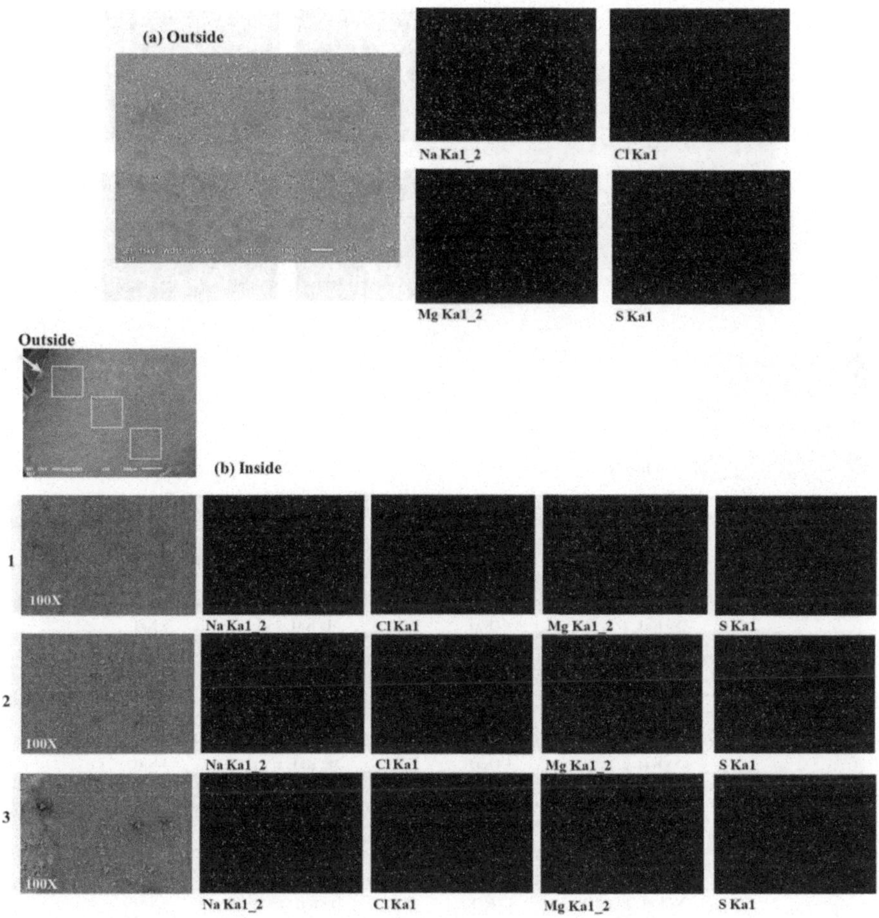

Fig. 5. SEM-EDX micrographs of (a) outside, (b) inside of FA sample soaked in the salt solution for 14 days (1,2 and 3 were analyzed from outside to inside of sample, respectively).

compressive strength of geopolymer concrete immersed in salt water [10]. The presence of magnesium chloride may counteract the detrimental effects of magnesium sulfate, known as Ettringite formation, which can compromise concrete structures through sulfate corrosion [11]. Measurement of chloride penetration is crucial for assessing the extent of concrete damage by chloride ions, revealing a correlation between porosity and chloride penetration. Higher total porosity correlates with increased chloride penetration. Notably, geopolymer concrete demonstrates superior compressive strength compared to OPC in saltwater environments.

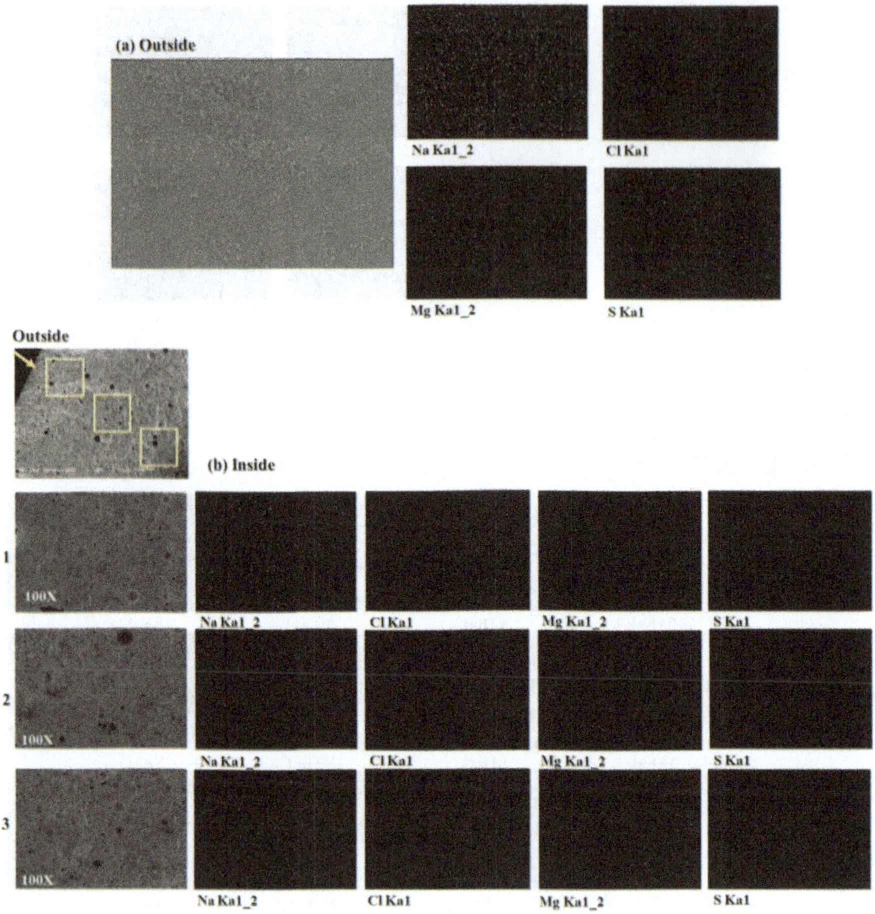

Fig. 6. SEM-EDX micrographs of (a) outside, (b) inside of FASiO2 sample soaked in the salt solution for 14 days (1,2 and 3 were analyzed from outside to inside of sample, respectively)

4 Conclusion

This study aimed to study the chloride and sulfate resistance of geopolymer based on waste alternative materials which were fly ash and silica fume. The chemical and microstructural analysis were evaluated for understanding the mechanisms and chemical reactions of geopolymer under soaking in the high concentration salt solution as compared to those soaked in the water. The geopolymer of fly ash and the combination of fly ash and silica fume showed lower chloride and sulfate penetration than OPC type 1. It also indicated that the geopolymer with the combination of fly ash and silica fume provided the improvement of salt resistance. Moreover, the geopolymer in this study was lightweight, environmentally friendly environmental and cost-effective for consideration used in the alternative concrete.

Acknowledgment. This study acknowledge support provided by (1) Fundamental Fund (FF) from National Research Council of Thailand (NRCT), (2) Suranaree University of Technology (SUT), (3) Thailand Science Research and Innovation (TSRI), (4) National Science, Research and Innovation Fund (NSRF) (project code 90464) and finally, the authors would like to thank NWN trading and SIE-CON for raw materials (fly ash) support.

References

1. Davidovits, J.: Global warming impact on the cement and aggregates industries. World Resour. Rev. **6**(2), 263–278 (1994)
2. Zivica, V., Palou, M., Križma, M.: Geopolymer cements and their properties: A review. Build. Res. J. **61** (2015)
3. Amin, M., Abdelsalam, B.A.: Efficiency of rice husk ash and fly ash as reactivity materials in sustainable concrete. Sust. Environ. Res. **29**(30) (2019)
4. Sathawane, S.H., Vairagade, V.S., Kene, K.S.: Combine effect of rice husk ash and fly ash on concrete by 30% cement replacement. Proc. Eng. **51**, 35–44 (2013)
5. Singh, N.S., Thokchom, S., Debbarma, R.: Properties of fly ash and rice husk ash blended geopolymer with sodium aluminate as activator solution. Eng. Appl. Sci. Res. **48**(1), 92–101 (2021)
6. Shaikh Faiz, U.A.: Effects of alkali solutions on corrosion durability of geopolymer concrete. Adv. Concr. Constr. **2**(2), 109–123 (2014)
7. Olivia, M., Nikraz, H.: Properties of fly ash geopolymer concrete designed by Taguchi method. Mater. Des. **1980–2015**(36), 191–198 (2012)
8. Gunasekara, C., et al.: Chloride induced corrosion in different fly ash based geopolymer concretes. Constr. Build. Mater. **200**, 502–513 (2019)
9. Halim, L.N., Ekaputri, J.J.: And Triwulan, the influence of salt water on chloride penetration in geopolymer concrete. MATEC Web Conf. **97**, 01002 (2017)
10. Singh, N.B.: Fly ash-based geopolymer binder: A future construction material. Minerals **8**(7), 299 (2018)
11. Bikić, F.: Investigation of the influence of magnesium chloride to sulfate corrosion of concrete. Bull. Chem. Technol. Bosnia Herzegovina **46**, 7–12 (2016)

Powder Flow Properties and Hopper Design

Jan Baeyens[1], Miao Yang[1], Shuo Li[1], Zehao Li[2], Helei Liu[3], and Yimin Deng[3](✉)

[1] Beijing Advanced Innovation Centre for Soft Matter Science and Engineering, Beijing University of Chemical Technology, Beijing 100029, China
[2] School of Life Science and Technology, Beijing University of Chemical Technology, Beijing 100028, China
[3] School of Chemistry and Chemical Engineering, Beijing Institute of Technology, Beijing 100081, China
deng.ymin@gmail.com

Abstract. The handling and processing of powders can cause problems unless appropriate steps are taken when storing and transporting them - especially when made up of fine particles. The present chapter will consider the storage of powders in hoppers and their discharge. As far back as man has harvested and stored crops, people have stored particulate solids. Prior to 1960, the design of storage bins was an art. This changed when Andrew W. Jenike identified the criteria that affect material flow in storage bins. A proper hopper design is based on the flow properties of the bulk solid to be stored and a thorough understanding of these properties is mandatory.

Keywords: fine particles · hopper · particle size

1 Flow Properties of Bulk Solids in Storage Hoppers

To avoid flow problems in hoppers and bulk solids handling equipment (obstruction, segregation, irregular flow, flooding, etc.), it is necessary to determine the flow properties of powders. These properties depend on several parameters, e.g. particle size distribution, particle shape, chemical composition of the powder, moisture content and temperature. Several testing devices are available to determine the flow properties of bulk solids thus making it possible to systematically determine the flow behaviour in dependence of all different parameters.

The physical characteristics of bulk solids are insufficient to design transport and storage systems. These systems have to be investigated in view of the flow properties of the powders. It is commonly proposed that the angle of repose (AOR) determines flow properties. This is an oversimplification, as the angle of repose depends on the operation type, the equipment used and the product history, as shown in Fig. 1. The angle of repose is therefore not a dominant property, although a correlation with cohesiveness has been established [1, 2].

Jenike [3–5] and Johanson [6–8] introduced a thorough analysis of flow properties of bulk solids. The most important properties are given in Table 1, together with the

J. Baeyens et al. (Eds.): ESRE 2023, pp. 76–96, 2025.
https://doi.org/10.1007/978-981-96-4345-5_7

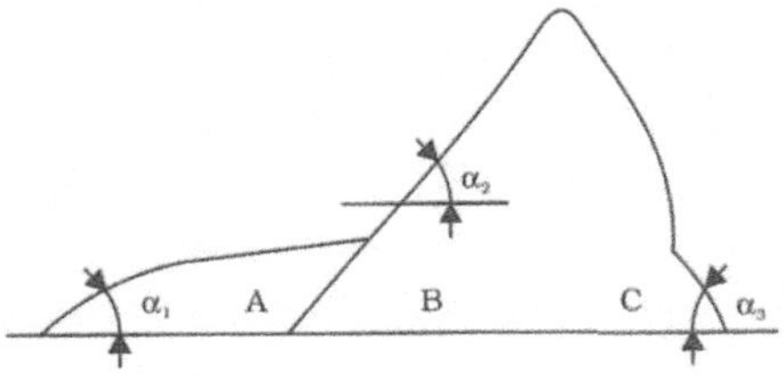

Fig. 1. The angle of repose depends on the way of unloading: (A) AOR after aeration, (B) AOR before aeration, (C) AOR as part of the powder is removed from B.

Table 1. Flow properties and the influence of moisture content and compaction.

Flow property	Unit	Description	Influence of increasing moisture content	Influence of increasing compaction stress
angle of internal friction, δ	°	kinetic friction during constant flow	↑	↑
angle of internal friction, φ	°	friction when powder starts to flow	↓	↑
angle of wall friction, φ_W	°	$\tan(\varphi_W)$ = coefficient of kinetic friction between powder and wall	↑ or ↓	
bulk density, ρ_B	kg/m^3	/	↓	↑
shear stress, f	kg/m^2	measure of cohesiveness	↑	↑
Permeability, k	m/s	superficial gas velocity through the powder at ρ_B	↑	↓

influence of moisture content and compaction, and are illustrated in Fig. 2 for the flow of polyethylene powder in a galvanized steel hopper [1].

The effect of internal hopper liner s [9] and of fitting hoppers with inserts [10] can positively alter this overall behaviour. Similar results were also presented by Komuczki et al. [11] and by Ali et al. [12].

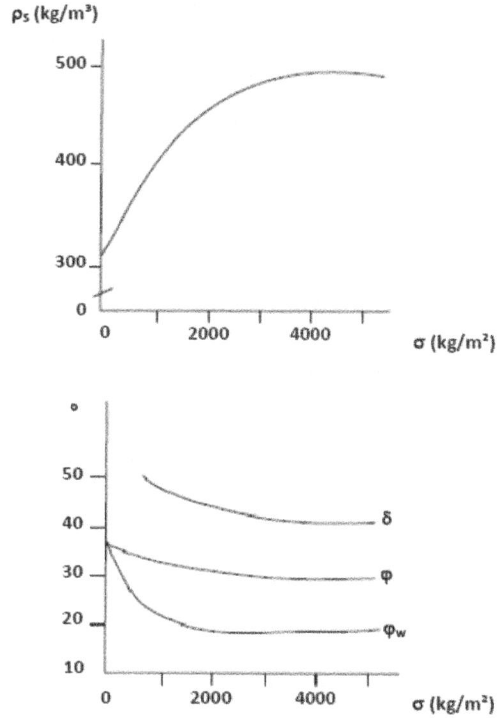

Fig. 2. Flow properties of polyethylene powder at increasing compaction stress σ.

2 Basic Principles

2.1 Relevant Parameters that Influence Flow Properties

Geldart [13] described the behaviour of a bed of powders when fluidized with air, and divided them into four distinct groups, according to recognizable features.

- Group A powders exhibit dense phase expansion after minimum fluidization and prior to the commencement of bubbling,
- Group B powders bubble at the minimum fluidization velocity,
- Group C powders are difficult to fluidise,
- Group D powders can form stable spouted beds.

Particle density and the average *particle size* determine the group classification. Powders in groups A, B and D rarely encounter problems during discharge from a storage hopper. Group C powders on the other hand are cohesive, which induces the risk of arch formation or rat-holing. The delineation of the C/A boundary was improved by Kong et al. [14] and by Li et al. [15].

Apart from particle size and density, there are three other parameters of importance: moisture content, temperature, particle sphericity [16] and the possible presence of flow-enhancing additives. A high *moisture content* will increase cohesion through H_2O bridges: to maintain powders flowing and to avoid arch formation, the powder can be dried before storage, the hopper can be aerated or the hopper walls can be heated. The influence of *temperature* on the flow properties of a powder is most relevant in the food or pharmaceutical industry and when processing bulk solids with a low melting point. By adding certain *additives* to a cohesive powder, the flow properties can be improved. Using additives implies however a certain and possibly unacceptable contamination of the bulk solid.

2.2 Storage Hoppers

In general, a hopper is used for storage since it holds a reserve of material that can be fed into the process at a required rate and at a desired time. It is essential to ensure that the material will gravitationally discharge when required, and that this flow will not be interrupted as long as the hopper is not empty. The design method for uninterrupted discharge will be discussed further in this chapter and is applicable for most materials. When however a powder is too cohesive and will not flow from a hopper of acceptable design, external flow enhancement methods are necessary.

Based on the flow patterns occurring in storage hoppers during their discharge, two main types can be distinguished: core flow (or funnel flow) and mass flow hoppers. These two flow modes are shown in Fig. 3, representing a cross section through a hopper filled with a bulk solid.

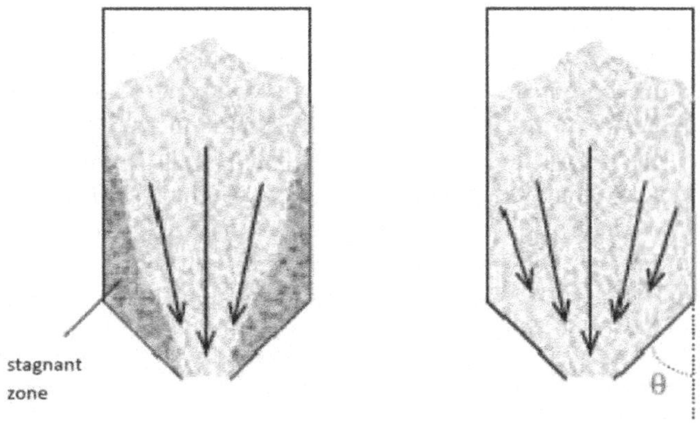

Fig. 3. Core (or funnel) flow (left) and mass flow (right) in a hopper with angle θ [3].

When a small amount of particles is discharged from a *core flow hopper*, a distinct region can be defined where downward motion of the particles takes place. In the middle, the top surface begins to fall. As time proceeds, the core will widen, though still well-defined. Particles outside the core (near the wall) remain stagnant: even when the hopper is almost empty, there will still be some material undisturbed. A core flow hopper is therefore characterized by "dead zones" when discharged. For a cohesive material, flow from the hopper can stop, due to both the formation of an arch between the hopper walls, strong enough to resist the static pressure head of the remaining load; or due to piping or ratholing, where the material directly above the outlet of the hopper drops out, leaving an empty cylinder with almost vertical powder walls.

Another disadvantage is that the material discharged from the hopper can vary significantly in rate and bulk density, even when stoppage does not occur. When a mixture of different particle sizes is fed into a core flow hopper, the mixing quality can be lost and segregation may occur, with resulting variation in bulk density.

If an arch is formed above the hopper outlet, it can collapse, exerting very large stresses on the lower part of the hopper, with possibly disastrous effects. When arch collapsing inside a hopper happens with very aeratable powders, the whole content of the hopper is fluidized, which results in a very rapid discharge. This condition is referred to as *flooding* and can be very dangerous: the factory floor gets covered with powder, or someone might get buried underneath. The problem of flooding, and also other disadvantages of core flow hoppers, can be resolved by modifying the hopper into a mass flow type.

In a *mass flow hopper*, every particle is in motion during discharge. There is however a velocity gradient: particles near the centre of the hopper move slightly faster than those near the wall. The definition of a mass flow hoper therefore does not require that all particles move with the same velocity but that they all move: this type of storage hopper is characterized by the absence of dead zones.

The use of mass flow hoppers comes with different *advantages*. During discharge, the stress distribution is such that a stable arch or pipe cannot occur: the flow will not stop and problems related to arch collapsing inside the hopper will not occur. The rate of discharge and the bulk density will be much more consistent. As arch forming is unlikely, flooding is excluded. A mass flow hopper reduces segregation. Despite these advantages, not all hoppers should be of the mass flow type, as for several applications, a core flow hopper will work perfectly and for example in the storage of free-flowing material without size segregation or undesirable effects. Furthermore, when a hopper is to be erected in an existing building, the head room can be insufficient for a mass flow hopper, as it will be taller than a core flow hopper for a given capacity. Finally, particles sliding along the mass flow hopper wall can lead to transfer of material between the particles and the wall. If this creates serious problems (nuclear fuels contaminated with even the tiniest impurities, storage of very abrasive particles, etc.), the mass flow hopper should be replaced by a core flow type.

3 Consolidation of Bulk Solids

The flowability of a bulk solid is affected by the adhesive forces between individual particles. With dry bulk solids, Van der Waals interactions play an essential role, whereas liquid bridges between the particles are usually more important in moist bulk solid. Both types of adhesive forces however depend on the distance between particles and the particle size [15].

Some bulk solids continue to gain strength when stored for a long period under compressive stress: this effect is called *consolidation*. Possible mechanisms for this consolidation are due to the effects of adhesive forces [17].

- Solid bridges formed by solid crystallising when drying moist bulk solids, where the liquid phase is a solution of a solid and a solvent;
- Solid bridges from the particle material itself, e.g. after some material at the contact points has been dissolved by moisture;
- Bridges formed by sintering during storage at high temperatures approaching the melting temperature;
- Chemical or biological processes.

Apart from these adhesive forces, other forces acting on the bulk solid also influence the flowability. The influence of adhesive forces on flow behaviour increases with decreasing particle size, in line with their cohesive nature.

A compressive force acting from outside on a bulk solid element can increase the adhesive forces: particles are pressed against each other leading to plastic deformation at the points of contact. This results in an increase of the contact areas, a closer particle approach and a strengthening of the adhesive forces. This dependence of the adhesive forces on external forces is characteristic of bulk solids, especially for cohesive ones. It is hence always necessary to take the stress history (forces or stresses previously acting) of a bulk solid into account.

When certain materials are left undisturbed in a hopper, they gain in strength due to gravity. This can influence discharge, even when the mass flow hopper has been correctly designed. This *time consolidation* or caking can be measured in a shear cell, and the hopper design can be adjusted.

4 Hopper Design

As stated before, mass flow hoppers are the preferred storage mode due to the obvious advantages. Whether a hopper is of the mass or core flow type depends on the flow properties of the particulate material stored: the properties of this material should be first defined and measured. Thereafter the Jenike design method can be applied.

4.1 Stresses in Bulk Solids

A bulk solid element in a container is shown in Fig. 4. In the vertical direction, a positive normal compressive stress ($\sigma_{\mathrm{v}} > 0$) is exerted on the bulk solid. The horizontal stress (σ_{h}) is a result of the vertical stress and it is smaller than the vertical stress exerted on the bulk solid from the top. The ratio of horizontal versus vertical stress is defined as the stress ratio (K), and typical values lie between 0.3 and 0.6 [18].

In a bulk solid, different stresses can be found in different cutting planes other than the vertical and horizontal, which can be analysed by using a simple equilibrium of forces. Assuming that no shear stresses are exerted on the top or bottom surface of the bulk solid element, the shear stresses in these planes are equal to zero. Furthermore, the lateral walls are assumed frictionless, so no shear stresses are acting on the lateral walls either. Thus only the normal stresses shown in Fig. 4 are acting on the bulk solid element from the outside. Figure 5 shows a volume element with triangular cross-section cut from the bulk solid element. Using a simple equilibrium of forces at this volume

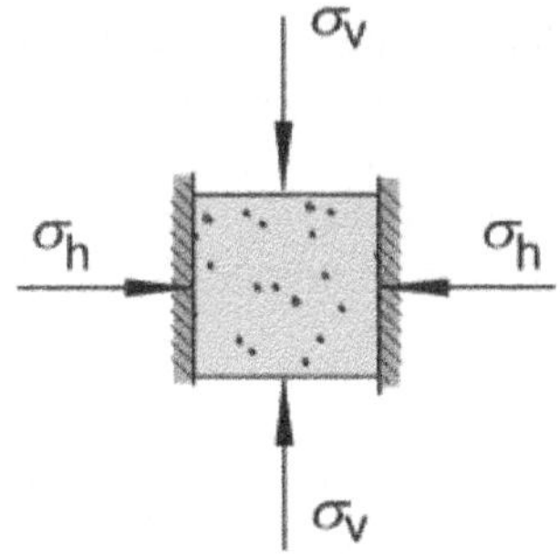

Fig. 4. Horizontal and vertical stress in a bulk solid element.

element, the normal stress (σ_α) and the shear stress (τ_α) - acting on a plane inclined by an arbitrary angle (α) - can be calculated:

$$\sigma_\alpha = \frac{\sigma_v + \sigma_h}{2} + \frac{\sigma_v - \sigma_h}{2}\cos(2\alpha) \tag{1}$$

$$\tau_\alpha = \frac{\sigma_v - \sigma_h}{2}\sin(2\alpha) \tag{2}$$

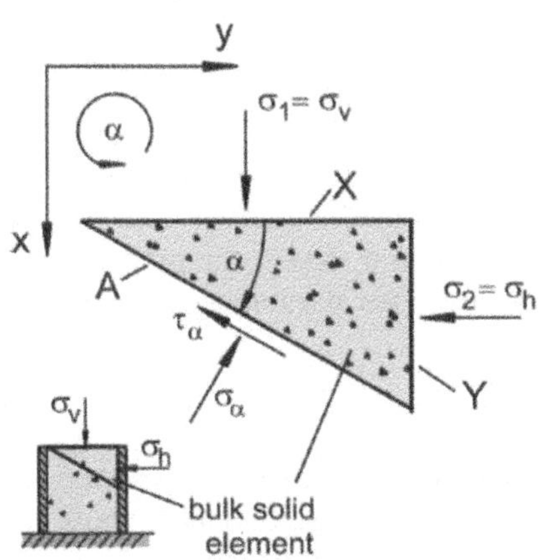

Fig. 5. Force equilibrium on a bulk solid element.

The pairs of values (σ_α,τ_α) for all possible angles can be plotted in a σ,τ-diagram, which results in a circle. This so-called *Mohr stress circle* represents the stresses in all possible cutting planes within a bulk solid element, and is shown in Fig. 6. The centre of the circle is located at $\sigma_m = (\sigma_v + \sigma_h)/2$ and $\tau_m = 0$; and the radius is $\sigma_m = (\sigma_v - \sigma_h)/2$. Each Mohr stress circle has 2 points of intersection with the σ-axis: the normal stresses defined through these points are called principal stresses (σ_1 the major and σ_2 the minor principal stress). When both principal stresses are known, the Mohr stress circle is totally defined [19].

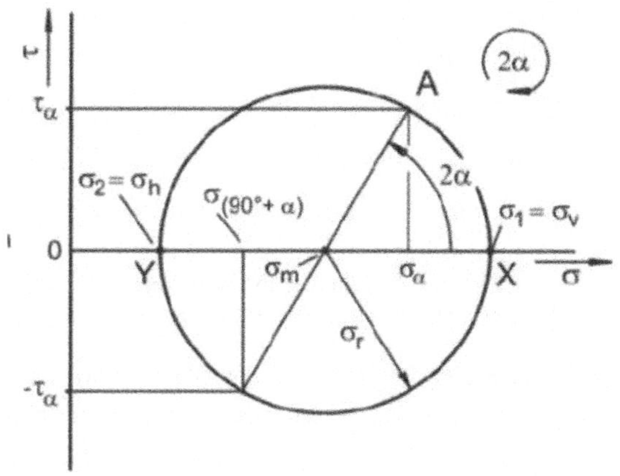

Fig. 6. Mohr stress circle.

It can be concluded that the state of stress in a bulk solid cannot be completely described by a single numerical value. Depending on the actual load acting on a bulk solid element, the corresponding Mohr stress circles will have different radii, centres and principal stresses.

4.2 Uniaxial Compression Test

A bulk solid flows easily if it does not consolidate excessively and and will discharge from a silo or a hopper due to gravity forces alone without requiring flow promotion (aeration, vibration, etc.). Products are poorly flowing if they experience flow obstructions or consolidate during storage or transport. To make a quantitative statement on *flowability*, an objective characteristic value is required to take into account those physical characteristics of the bulk solid that are responsible for its flow behaviour.

Flowing means that a bulk solid is affected due to the loads acting on it: the magnitude of this load is a measure of flowability, and it can e.g. be measured with the *uniaxial compression test*, as shown in Fig. 7. A hollow cylinder is filled with a fine-grained bulk solid, and the wall is assumed frictionless. The bulk solid is then submitted to consolidation stress σ_1 in the vertical direction. The more the volume of the bulk solid is reduced, the more compressible it is. After consolidation, the consolidation stress is relieved and the hollow cylinder is removed. The consolidated cylindrical bulk solid is thereafter loaded with an increasing vertical compression stress or unconfined yield strength σ_c: at a certain stress, the specimen will break (fail).

In the uniaxial compression test, the force of gravity can be neglected and no friction between the wall of the hollow cylinder and the bulk solid can be assumed. Within the bulk solid specimen, both vertical (σ_v) and horizontal (σ_h) stresses are thus constant. Therefore at each position in the sample, the state of the stress - represented by a Mohr stress circle - is identical. During consolidation, the vertical normal stress (σ_1) acts on top of the specimen. Perpendicular to this stress, a lesser horizontal stress prevails

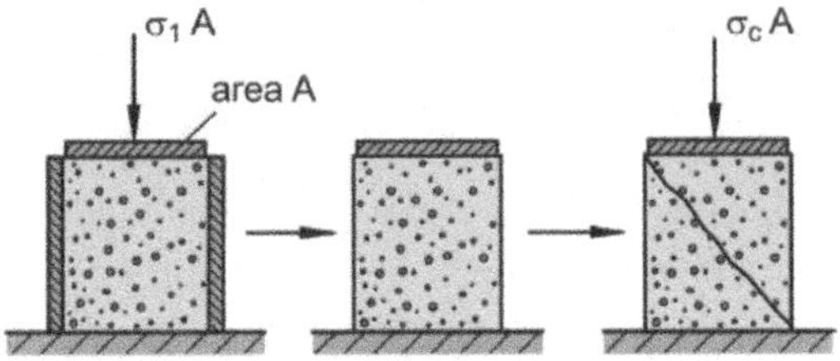

Fig. 7. Uniaxial compression test.

according to the stress ratio (K). Nowhere in the sample shear stresses will be found, so $\tau = 0$. Figure 8 shows the pairs of values (σ,τ) for different horizontal and vertical cutting planes. Both points are located on the σ-axis because $\tau = 0$. The Mohr stress circle A is thus well defined and describes the stresses in the bulk solid sample at consolidation. The vertical stress is identical with the major principal stress, since in the vertical plane the shear stress is zero. The minor principal stress is equal to the horizontal stress. After the sample has been relieved of the consolidation stress and the hollow cylinder has been removed, it is loaded with increasing vertical stresses. The vertical and horizontal stresses are principal stresses. The horizontal stress is independent of the vertical load and always equal to zero, since the lateral surface of the specimen is uncovered and not loaded. During the increasing vertical load, the stress states at different load steps are represented by stress circles with increasing diameter (B_1, B_2 and B_3). When the specimen fails, the stresses in the bulk solid sample are represented by the Mohr stress circle B_3. The corresponding load causes incipient flow of the specimen, so the yield limit of the bulk solid must have been attained in one cutting plane. Figure 8 shows a possible yield limit, reached by the Mohr stress circle B_3. The real course of the yield limit can however not be determined with the uniaxial compression test only. If in addition to the vertical stress also a constant horizontal stress is applied on the specimen, one would likewise find stress circles that indicate failure of the specimen and reach the yield limit, e.g. stress circle C. The yield limit can therefore be considered the envelope of all stress circles that indicate failure of a bulk solid sample [3, 18].

4.3 Time Consolidation

When bulk solids are stored for a longer time at rest under a compressive stress, they can increase in strength: this effect is called time consolidation or caking, as already mentioned. The same test as described above can be used to determine the time consolidation, but the consolidation stress has to be applied for a longer period. The resulting pairs of values (σ_1, σ_c) form the so-called *flow function*. Figure 9 shows the flow functions for different consolidation times: in this example, the unconfined yield strength increases with increasing storage time. This result is true for many bulk solids, but not for all. Some bulk solids undergo none or only very slight consolidation over time, whereas others undergo a large increase in unconfined yield strength after storage periods of only a few hours. It does not increase for longer storage periods. These differences are due to different physical, chemical or biological effects that cause consolidation over time.

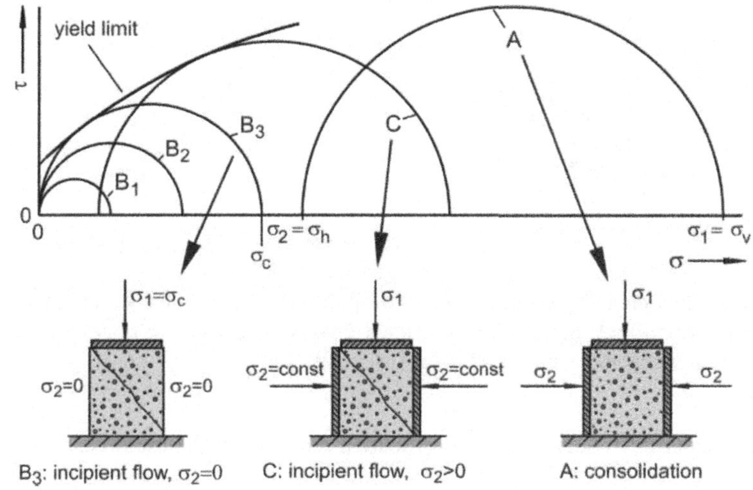

Fig. 8. Measurement of unconfined yield strength in a σ, τ-diagram [18].

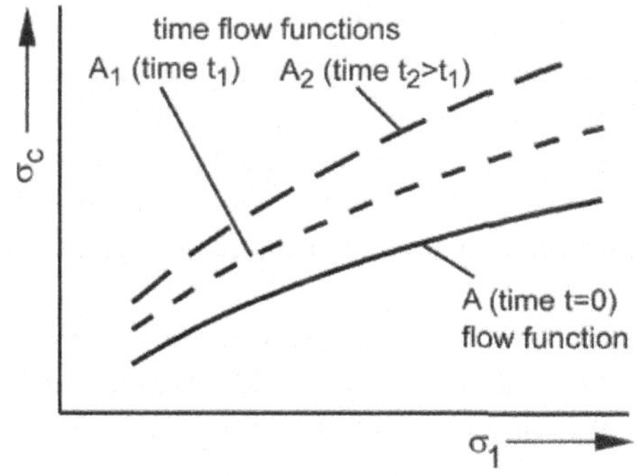

Fig. 9. Flow functions for different storage times: t < t1 < t2.

Flowability of a bulk solid is characterized mainly by its unconfined yield strength, in function of the consolidation stress and the storage period. With the ratio of σ_1/σ_c defined as the flow function ff_c, flowability can be numerically defined. The larger ff_c, the better a bulk solid will flow. Using this ratio, flow behaviour can be defined as follows (Fig. 10).

If the bulk solid shows a time consolidation effect, the unconfined yield strength will increase with increasing storage period. This will result in a decrease in flowability with time. If a bulk solid gains strength with an increasing period of storage at rest at a certain consolidation stress, it will be more difficult to get this bulk solid to flow: flowability will hence decrease with increasing storage period, as shown in Fig. 10.

It can be concluded that it is very important to test a bulk solid at defined and known conditions. The flow behaviour of different bulk solids can only be compared if

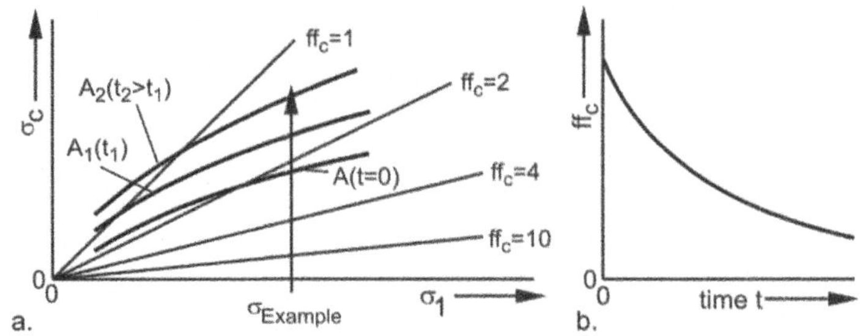

Fig. 10. Flow function and lines of constant flowability (left) and influence of storage time on flowability (right) ($ff_c < 1$: not flowing; $1 < ff_c < 2$: very cohesive; $2 < ff_c < 4$: cohesive; $4 < ff_c < 10$: easy-flowing; $10 < ff_c$: free-flowing) (Schulze, 2021).

all measurements have been performed for identical consolidation stresses and storage periods.

4.4 Shear Testing

To ensure that a powder will slide down the hopper cone during discharge, a minimum cone slope is necessary. Powder flowing is a result of the application of shear stress, and the magnitude of this stress depends on the bulk density of the material, which in turn depends on the compacting stress to which it has been exposed and the normal stress acting on the slip plane. Characterizing the *flowability* of a powder therefore comprises investigating the relationship between shear strength and compacting stress: this is tested in a so-called shear cell.

When a bulk solid is fine-grained and cohesive, the use of the uniaxial compression test is problematic: the obtained unconfined yield strength values will be too low, and preparation of the hollow cylinder to obtain frictionless walls is very tedious. Moreover, other important parameters cannot be determined with this test. Flow properties of fine-grained bulk solids are therefore measured with so-called shear testers. For coarse-grained bulk solids however, it is an appropriate technique to measure the time consolidation [18].

Figure 11 shows the principle of a shear test. In a shear cell, a bulk solid specimen is subject to a vertical charge, the normal stress Shear deformation is applied by moving the top plate with a constant velocity v, which results in a horizontal shear stress τ. The resultant force F_R will increase with increasing shear stress.

In this test, the bulk solid is first consolidated: this step is called *pre-shear*. The specimen is loaded vertically by a normal stress σ_{pre} and then the specimen is sheared. The shear stress τ will first increase with time, then the curve will flatten and finally remain constant: this constant shear stress is called τ_{pre}. Starting from this point, the bulk density will remain constant and the flow that will occur is called steady-state flow. This first step of the shear test results in a first point of the yield locus: (σ_{pre}, τ_{pre}) or the so-called pre-shear point. At this point, the specimen in the cell is defined as critically consolidated. The second step of the shear test is called *shear* or *shear to failure*. The normal stress acting on the specimen is decreased to a value $\sigma_{sh} < \sigma_{pre}$.

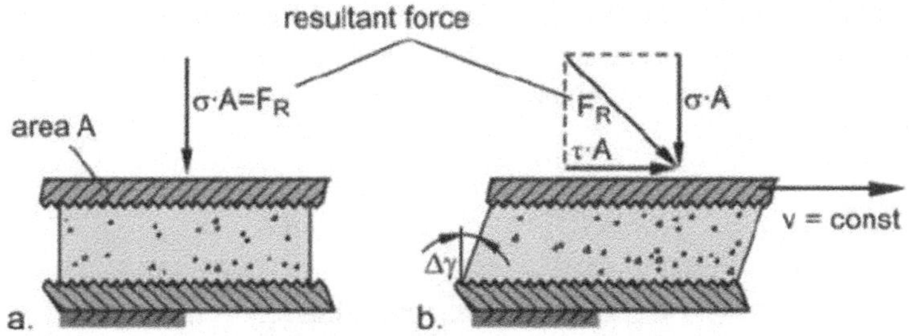

Fig. 11. Bulk solid specimen with (a) initial loading with normal stress and (b) shear deformation (constant velocity v).

When a sufficiently large shear stress is attained, the specimen will start to flow (fail), as particles will start to move. The material will start to dilate, decreasing the bulk density and the shear stress. The maximum shear stress indicates incipient flow and the corresponding pair of values (σ_{sh}, τ_{sh}) is a point of the yield locus. When several of these tests are performed, with shearing to failure under different normal stresses but at identical pre-shear normal stresses (so that each specimen reaches the same state of consolidation), the course of the yield locus can be determined. Figure 12 shows the different steps in determining this yield locus.

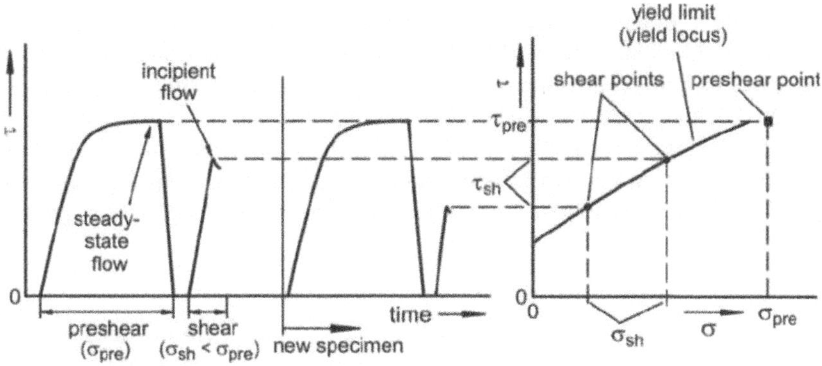

Fig. 12. Plot of shear stress versus time and yield locus [18].

4.5 Jenike Shear Cell

The most common used shear tester is a cell introduced by Jenike in 1960 [3]. The circular cell has a diameter of 93 mm and consists of a bottom ring, an upper ring with the same diameter and a lid, as shown in Fig. 13. Both rings are filled with the powder to be tested and the lid is placed on top. By putting weight on the lid, a normal force F_N is exerted on the specimen and the upper ring is displaced horizontally. The shear force F_S exerted is measured by a load cell. The displacement of the upper ring results in a shear deformation of the bulk solid. By dividing the normal force F_N and the shear force F_S by the cross-sectional area of the cell A, the normal stress σ and the shear stress τ acting in the horizontal plane between upper and bottom ring are determined.

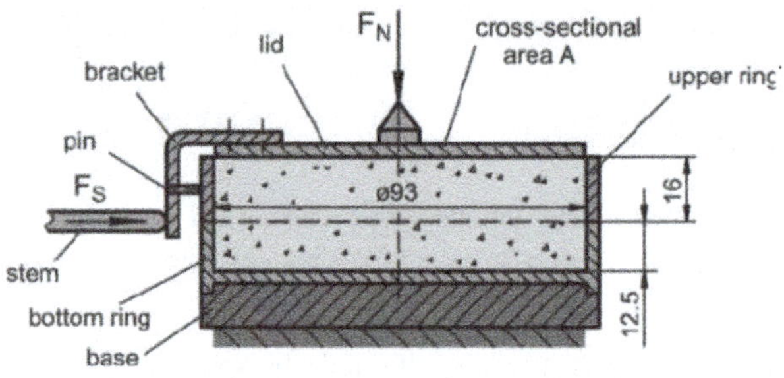

Fig. 13. Jenike shear cell.

The shear strength of a sample depends on its bulk density, determined by the compacting force to which it has been subjected. When this force is removed, the sample retains its bulk density. When a vertical force is then applied to the sample and the ring, and its contents are pushed forward with a constant speed, the volume of the sample will generally change (expansion or contraction, depending on the normal load). By applying different normal loads, it can be found under which load the sample will not change in volume: this condition is defined as the critical state. In the next series of tests, different normal loads, all less than the critical one, are applied and the corresponding maximum shear stresses to initiate movement are measured. These test results are then plotted in the so-called Jenike yield locus, shown in Fig. 14: the shear stress needed to initiate movement versus the normal stress acting on the shear plane. A Jenike yield locus represents the results of a series of tests on samples with the same initial bulk density. By changing this density, more loci can be obtained, which is necessary for hopper design using the Jenike method.

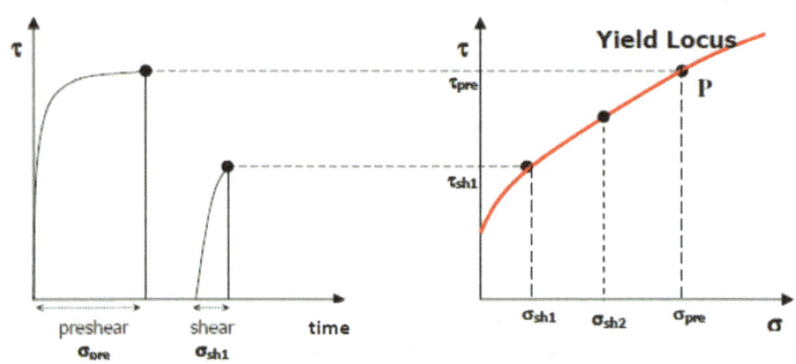

Fig. 14. Yield locus.

4.6 Yield Locus

The flow properties of a powder can be determined from the yield locus, as shown in Fig. 15. First the Mohr stress circle is drawn tangential to the yield locus and intersecting at the point of steady-state flow. This circle represents the stresses in the specimen at the end of consolidation and the major principal stress of this circle is equal to the relevant consolidation stress σ_1. Secondly a stress circle is drawn tangential to the yield locus and through the origin. The major principal stress of this circle is equal to the unconfined yield strength σ_c. Finally a straight (dotted) line is drawn through the origin and tangent to the greater Mohr circle: this represents the effective yield locus. The angle φ_e represents the effective angle of internal friction and can be regarded as a measure of the internal friction at steady-state flow.

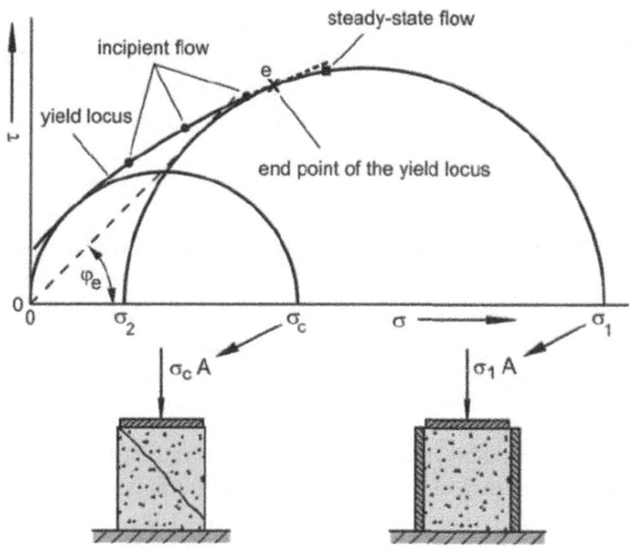

Fig. 15. Yield locus.

4.7 Jenike Design Method

For designing mass flow hoppers using the Jenike method, the concepts of yield locus and Mohr circle are brought together: both graphs can be drawn on the same diagram, as shown in Fig. 16. The Jenike yield locus represents the stresses needed to initiate flow in a powder at a density at which the yield locus was obtained, whereas any set of stresses applied to the material can be represented by a Mohr circle. If the Mohr circle touches the yield locus, the circle represents a set of stresses at which powder will flow. If however the circle lies wholly below the yield locus, the powder will not start flowing under these stress conditions.

In the above diagram, numerous Mohr circles can be drawn. There are however two circles of particular interest:

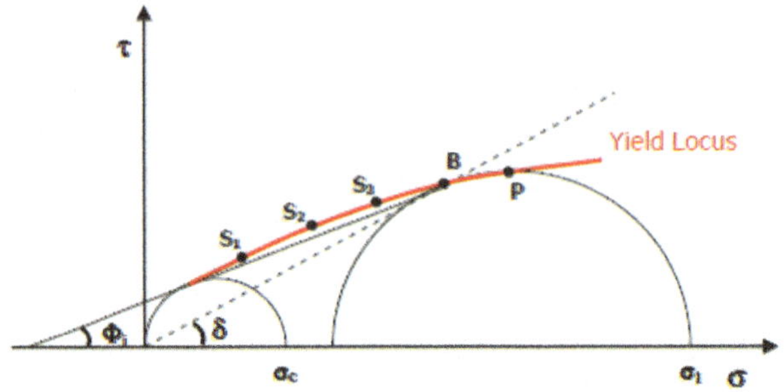

Fig. 16. Jenike design method [19].

- When arch forming occurs at the outlet of a hopper, flow will stop. An element of the material in the surface of the arch will lie in a free surface. This surface will have no normal stress and no shear stress acting on it, so the Mohr circle for this element will pass through the origin of the graph. A circle passing through the origin and touching the yield locus hence represents the stress needed to initiate flow in the surface of the arch, causing it to collapse. The diameter of that circle is a measure of the major principal stress needed to make the arch collapse, referred to as the *unconfined yield stress* σ_c.
- Each Jenike yield locus has an end point, which represents the conditions for failure without change in volume, and also the stress conditions in the powder as it is flowing downwards towards the outlet. The Mohr circle that touches the yield locus tangentially at this end point, represents the stresses acting on the powder during arch forming. The diameter of this Mohr circle is a measure of the *major principal compacting stress* σ_1.

Each yield locus will result in a pair of values (σ_1, σ_c). If these values are plotted, a graph is formed, depending on the cohesiveness of the material tested. This graph is called the *material flow function* and represents the stress needed for arch collapsing in function of the compacting stress under which it was formed.

The second property obtained from the yield loci is the *effective angle of internal friction* δ of the powder, defined as the slope of a straight line drawn from the origin tangential to the end point Mohr circle.

The last property needed to design a mass flow hopper, is the *angle of wall friction between the particulate material and the wall material* ϕ_w. This property is necessary in the calculations, as it is important that particles are able to slide down the hopper wall. This angle can be obtained by replacing the base of the Jenike shear cell by a plate of the hopper construction material. The ring from the shear cell is placed on the plate, filled with powder and submitted to a shear force: the force needed to push the ring and its content forward at a constant speed and for different vertical loads on the lid is measured. The slope of the graph of these different shear forces against normal forces gives the angle ϕ_w.

After the characterisation of the flowability of the particulate material, the stresses inside the hopper have to be predicted. A small element in a bed of particulate material will be subjected to external stresses which will have two effects. Firstly the element will be compacted, which influences the bulk density and hence the flowability of the material. The higher the compacting stresses, the higher the shear stresses needed to cause the material to flow. While the compacting forces are being applied, shear stresses will be set up in the material which will attempt to make it flow. For any given element inside a mass flow hopper, the ratio of the compacting stress to the shear stress is constant, as shown by Jenike. This ratio is called the flow factor. The value of this factor depends on the effective angle of internal friction of the particulate material, the slope of the hopper wall, and the angle of friction between the powder and the hopper wall. The flow factor is neither a property of the powder, nor of the hopper: it depends on the whole system of a powder in a hopper. Values are obtained from a series of charts provided by Jenike [3], as shown in Fig. 17 for an angle of internal friction of 40°. These charts can only be used in the triangular region where mass flow is possible.

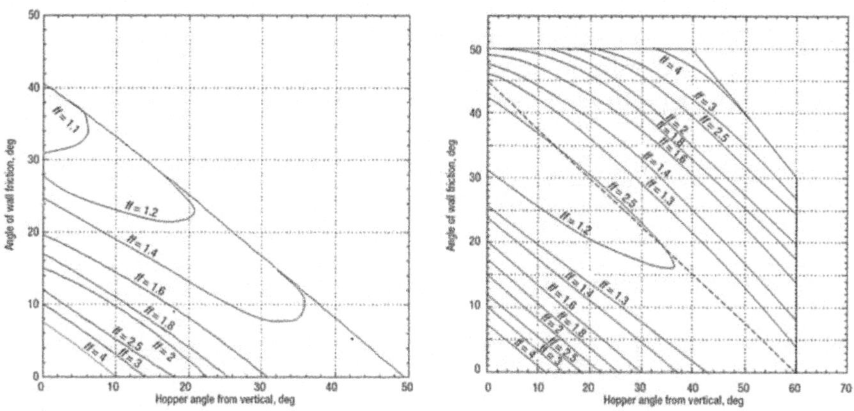

Fig. 17. Jenike chart at angle of internal friction: angle of wall friction, θ_w, versus angle of hopper cone, θ.

In the third stage of the hopper design, a comparison is made between the stress exerted on the powder inside the hopper and the stress needed to make the material flow. The aim is to design a mass flow hopper in such a way that the stress conditions for failure will be satisfied at every point inside the hopper.

The material flow function and the flow factor can be plotted in the same diagram, as the horizontal axis represents the consolidating stress for both graphs: this is illustrated in Fig. 18. At lower stresses, the flow factor lies below the material flow function: the stress applied to the powder is less than its shear strength, so arches or pipes can occur which results in stoppages. At higher stresses, the applied stress exceeds the shear strength: stoppages are not possible, so uninterrupted flow will occur. At the point where both graphs intersect, the applied stress is equal to the shear strength: this is called the critical applied stress CASS.

The critical size of the hopper outlet depends on the value of CASS, and Eq. (3) determines the circular outlet diameter D:

$$D = H(\theta)CASS/\rho_{\text{bulk}} \tag{3}$$

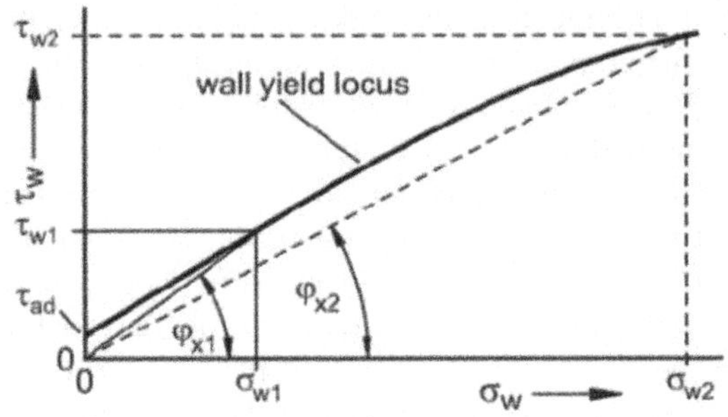

Fig. 18. Determination of the critical applied stress CASS.

with ρ_{bulk} the bulk density of the powder, and H(θ) a factor which takes into account that a part of the powder weight is supported by the hopper wall an does not contribute to the stresses in the powder near the outlet.

For different geometries however, *H(θ)* has to be adjusted, according to the values proposed in Fig. 19.

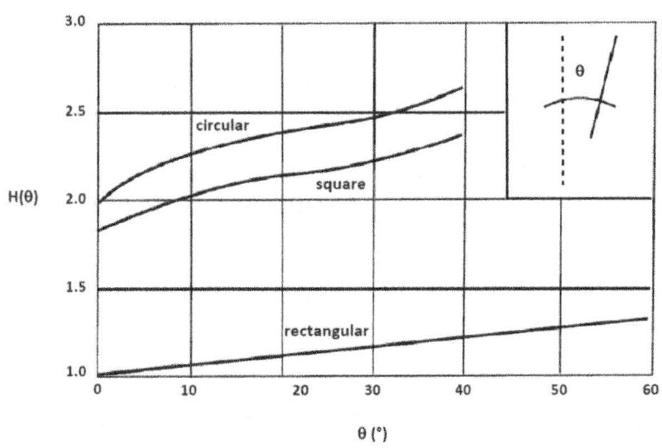

Fig. 19. H as a function of the outlet geometry: (1) circular, (2) square, (3) rectangular (L ≥ 3B).

As soon as the outlet dimensions are determined, it is necessary to control whether these dimensions are sufficient to discharge the requested amount of powder. This amount can be calculated using the formula of [21]:

$$g = \frac{4.V_s^2.\sin(\theta)}{D} + \frac{15.\rho_f^{\frac{1}{3}}.\mu^{\frac{4}{3}}.V_s^{\frac{4}{3}}}{\rho_p.d_p^{\frac{5}{3}}} \tag{4}$$

with $V_s = \frac{4.M_{s,max}}{\rho_B.\pi \cdot D^2}$

$M_{s,max}$ = maximum discharge rate (kg/s) through the outlet.
D = hydraulic diameter (m) of the outlet.

ρ_B = bulk density of the powder (kg/m^3).
ρ_p = absolute density of the powder (kg/m^3).
g = gravity constant = 9.81 m/s^2.
d_p = mean particle size of the powder (m).
ρ_f = air density (kg/m^3).
μ = air viscosity (kg/m.s).
θ = half top angle (°).

4.8 External Influence on Flow Properties

When it is not possible to install a mass flow hopper, due to the size or the capacity, external means can help to avoid funnel flow [15]:

- Using a vibrating hopper base that avoids arch formation;
- Fluidization of the powder in the cone;
- Applying a conveying screw with a variable pitch for uniform extraction;
- Using a conveying belt, a vibrating chute, a rotary valve, or a table extractor for uniform extraction from a rectangular outlet;
- Using external vibrators or internal vibrating rods.

4.9 Filling and Emptying Hoppers

A correct hopper design is only part of the good working conditions. As the method of filling or emptying a hopper has an influence on discharge, these parts of the system are also important.

When material is dropped from a considerable height into the hopper, it can be compacted very tightly near the hopper outlet, having more strength than assumed from the shear cell test. There are several solutions to avoid this problem: by dropping the powder against the wall of the hopper, by maintaining a small discharge during filling, etc.

In most cases, a hopper will be emptied by an extractor at a controlled rate. When a horizontal belt conveyor is placed under a slot outlet hopper, the belt will fill up with material falling onto the back end of the belt. As this material moves along with the belt, it prevents any more material falling out of the hopper onto the belt in the forward end of the slot. This results in an effective opening that is smaller than the design opening, and possibly smaller than required for mass flow: this creates an arch, which prevents further discharge. To prevent this, the feeder has got to have an increasing capacity in the direction of flow, so that it is never full and can take material from every part of the outlet. This can be achieved by sloping the belt downwards, by sloping the bottom of the hopper upwards or by increasing the width of the hopper outlet in the direction of the belt movement.

4.10 Other Measurement Methods for the Determination of Bulk Solid Properties

There are several other bulk solid properties that can easily be assessed. To determine the *angle of repose* α_M, different results are obtained from different simple test procedures, as shown in Fig. 20: for a conical heap (a), the discharge from a container with a central outlet (b), and using a rotating drum (c) [18].

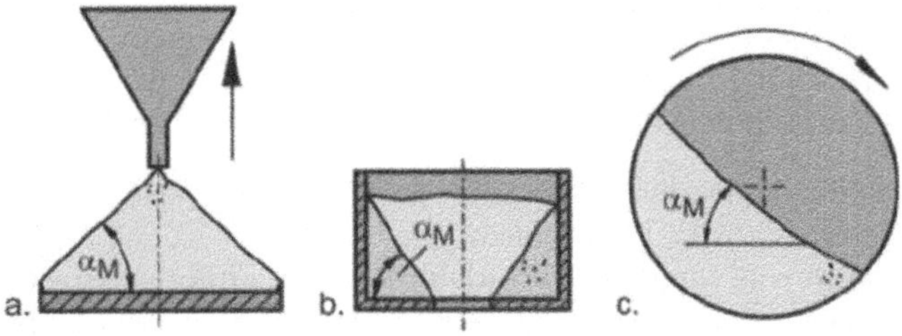

Fig. 20. Different methods to measure the angle of repose α_M [19].

Other simple test methods are the determination of the time required for a sample to discharge from a model silo; the determination of the compressibility by measuring the loose density and the tapped density (Hausner ratio); and the Carr index. Different angles of repose are measured as a function of the compressibility and the particle size distribution.

Finally, it should be remembered that additional factors of influence on the hopper design have been reported in literature, including failures and design issues in concrete silos [22], the shape optimization of conical hoppers [23], or the use of intermediate bulk containers [24]. The application of finite element modeling of biomass hopper flow was recently presented by Yi et al. [25].

5 Conclusions

Whereas the design of powder storage hoppers was previously considered as an "art", a full characterization of the stored particulate material and the use of appropriate modelling approaches enables a better and reliable design and construction of the hoppers.

The paper provides insight in the relevant parameters that affect powder flow properties and their possible consolidation during storage.

The hopper design methods are dealt with in terms of the stresses in bulk solids and the corresponding testing by different methods, including the widely applied Jenike shear cell.

The critical hopper parameters of the cone angle and outlet diameter are determined from the CASS calculation. External effects on the flow properties and practical aspects of hopper construction conclude the treatment.

References

1. Baeyens, J., Van, P.V.: Fluidized bed incineration of sewage sludge: A strategy for the design of the incinerator and the future for incinerator ash utilization. J. Hazard. Mater. **37**, 179–190 (1994)
2. Geldart, A.D., Abdullah, B.E.C., Hassanpour, C.A., Nwoke, C.L.C., Wouters, D.I.: Characterization of powder flowability using measurement of angle of repose. China Particuol **4**, 4 (2006)
3. Jenike, A.W. : Flow properties of bulk solids. Proc. Am. Soc. Test. Mater. 1168–1190 (1960)
4. Jenike, A.W.: Gravity flow of bluk solids. University of Utah, Engineering experiment station Salt Lake City, Bulletin 108 (1961)
5. Jenike, A.W.: Gravity flow of bluk solids. University of Utah, Engineering experiment station Salt Lake City, Bulletin 123 (1964)
6. Johanson, J.R.: Stress and velocity fields in the gravity flow of bulk solids. J. Appl. Mech. **31**, 499 (1964)
7. Johanson, J.R. : Feeding. Chemical Engineering, Deskbook Issue, 75–83 (1969)
8. Johanson, J.R., Colijn, H.: New design criteria for hoppers and bins. Iron and Steel Engineer, 85–104 (1978)
9. Reddy, Y.V., Venkatesh, S.: Flow characteristic evaluation of iron ore for design of bunkers and chutes with different liner materials. Mater. Today Proc. **76**, 95–102 (2023)
10. Sun, D., Lu, H., Cao, J., Wu, Y., Guo, X., Gong, X.: Flow mechanisms and solid flow rate prediction of powders discharged from hoppers with an insert - ScienceDirect. Powder Technol. **367**, 277–284 (2020)
11. Komuczki, D., Hesse, N., Schmidt, J., Satzer, P.: A step closer to continuous buffer preparation from solids: Predicting powder compaction and how to prevent it. New Biotechnol. **76**, 98–105 (2023)
12. Ali, M.A., Shaari, N.S., Yamin, M., Sahar, A., Yusof, Y.A.: Investigating the compressibility and flowability of cocoa powders: A solution for chocolate industry. J. Food Process Eng **46**, e14368 (2023)
13. Geldart, D.: Types of gas fluidization - ScienceDirect. Powder Technol. **7**, 285–292 (1973)
14. Kong, W., Tan, T., Baeyens, J., Flamant, G., Zhang, H.: Bubbling and slugging of geldart group A powders in small diameter columns. Ind. Eng. Chem. Res. **56**, 4136–4144 (2017)
15. Li, S., Baeyens, J., Yang, M., Li, Z.H., Deng, Y.M., Zhang, H.L.: Vibro-fluidization of cohesive particles. Powder Technol. accepted for publication (2023)
16. Liu, S.D., Gan, J.Q., Zou, R.P., Yu, A.B., Zhou, Z.Y.: Wall stress analysis in an unsteady hopper flow with ellipsoidal particles. Powder Technol. Int. J. Sci. Technol. Wet Dry Particul. Syst. 361 (2020)
17. Gandia, R.M., et al.: Effect of the Hopper angle of a silo on the vertical stress at the cylinder-to-hopper transition. Agronomy **12**, 830 (2022)
18. Schulze, D. : Flow properties of bulk solids. Powders and Bulk solids: Behavior, characterization, storage and flow, 57–100 (2021)
19. Rhodes, M.J.: Introduction to particle technology. John Wiley & Sons (2008)
20. Baeyens, J., Geldart, D.: The storage and flow of powders. CPA, Amsterdam (2004)
21. Carleton, A.J.: The effect of fluid-drag forces on the discharge of free-flowing solids from hoppers. Powder Technol. **6**, 91–96 (1972)
22. Maraveas, C.: Concrete Silos: Failures, design issues and repair/strengthening methods. Appl. Sci. 10 (2020)
23. Huang, X.J., Zheng, Q.J., Yu, A.B., Yan, W.Y.: Shape optimization of conical hoppers to increase mass discharging rate. Powder Technol. **361**, 179–189 (2020)

24. Wikstrom, H., Remmelgas, J., Solin, S., Marucci, M., Tajarobi, P.: Powder flow from an intermediate bulk container – Discharge predictions and experimental evaluation. Int. J. Pharm. 120309 (2021)
25. Yi, H., Lanning, C.J., Dooley, J.H., Puri, V.M.: Finite element modeling of biomass hopper flow. Front. Energy Res. **11**, 1162627 (2023)

Clean Energy Technology and Energy Conservation

Optimal Energy Management of a Hybrid Powered Marine Outfall with Energy Recovery from Wastewater Discharge

Kanzumba Kusakana(✉) and Sandile Phillip Koko

Central University of Technology, Free State, Bloemfontein 9300, South Africa
{kkusakana,skoko}@cut.ac.za

Abstract. In this paper, a model is developed to optimally manage the power flows between the different components of a grid-connected hybrid energy system supplying a marine outfall wastewater treatment plant. The system is equipped with photovoltaic/battery system and an energy recovery system generating power using pressurized wastewater supplied by the treatment plant. For simulation purpose, a case study of a marine outfall system situated in Cape Town is used to reveal the dynamic of the proposed setup and model. The aim of the developed model is to lessen the grid energy costs which operates under the variable time of use tariff. This can be achievable by capitalizing on the generated renewable power from both the photovoltaic and energy recovery systems, while optimizing the battery storage power flow. Based on the simulation results, the established model revealed a daily energy cost saving of around 75.5% that can be achieved from the proposed marine outfall's operation, as compared to the use of utility grid as the only source of electricity used to meet the same load demand.

Keywords: Marine outfall · energy recovery system · hybrid renewable system · cost minimization

1 Introduction

Hydropower has been used for decades for bulk electricity generation when compared to other renewable energy sources [1]. It is also considered to be the most reliable, supplying close to 19% of the world's electricity need [2]. Due to its longer lifespan as well as low operation and maintenance costs, it has proved to offer low energy production cost. Renewable energy power generation have increased by up to 760 GW between 2004 and 2013, of which 37.5% is mainly generated from hydropower systems [3].

Different topologies of hydropower generation systems been developed and implemented. This includes hydrokinetic, tidal, wave, pumped hydro storage, impoundment or diversion systems [1, 4–8].

Recently, interests had grown regarding wastewater discharge from marine outfalls as a source of renewable energy. These marine outfalls use pipes to discharge pressurized wastewater, as a means of discharging the treated municipal or industrial wastewater

J. Baeyens et al. (Eds.): ESRE 2023, pp. 99–112, 2025.
https://doi.org/10.1007/978-981-96-4345-5_8

into the sea [9]. The excess energy in the pressurized water can be recovered to generate useful energy using the same principle as the one used in conduits hydropower systems [10–14].

In Ref [15], the authors have proposed an engineering technique of recovering unused kinetic energy from the wastewater marine outfall, in order to convert it into valuable electrical energy. They have also looked at the different potential sites in South Africa.

In Ref [16], the authors have modelled and simulated a wastewater energy recovery system, using a hydro-turbine and alternator connected at the end of the pipeline. The results have revealed that the wastewater of the marine outfall has the likelihood of generating electricity during discharging process.

In Ref [17], the authors have assessed the performance of the selected marine outfall wastewater energy recovery system in order to conduct a techno-economic analysis. The aim was to analyze the simple payback period, breakeven-point and lifecycle cost indicators.

All of the abovementioned studies, have successful shown that the energy recovered from marine outfall wastewater discharge can be successfully converted into electricity at a competitive cost of energy produced. However, there is a need to investigate how these energy recovery systems will optimally perform in a scheme, when combined with energy storage systems and other renewable energy sources as part on the marine outfall's energy supply architecture.

Therefore, in this paper, the grid energy costs which operates under the variable time of use (ToU) tariff. An optimal energy management model has been introduced to lessen the operation cost of the existing marine outfall supplied by grid-connected solar PV system with a battery storage as well as a wastewater energy recovery system. The developed model has been employed and tested to a case in South Africa and the simulation results were analyzed and compared to the case where the grid is used alone to supply the marine outfall's load demand. The results proved that the use of the developed model has led to grid energy cost savings as compared to a case whereby the utility grid is the only source of electricity used to meet the similar load demand.

2 Mathematical Modelling

2.1 System Architecture

In marine outfalls, electricity is primarily used for pumping as well as for supplying all the process used for the wastewater treatment prior of being discharged into the sea. For the marine outfall under consideration, energy is supplied by both the solar PV system and the utility grid. The system has also a hydroelectric generator used for energy recovery from the excess pressure in the pipe and convert it to electricity.

Figure 1 shows the layout of the proposed system as it consists of different power flows. It can be seen that the marine outfall is supplied by the power from the solar PV in priority, which is split between the pumping and the other electrician needs. Whenever there is a power deficit from the solar PV system, the battery power and/or the utility grid power are/is utilized to supply the pumping as well as the other load. The cost incurred from utility grid power consumption must be minimized for economic purposes. The power recovered using the energy recovery generator can be used for recharging the battery storage system, for supplying both the pumping system and the other electrical equipment from the marine outfall.

At any point in time, the operation decision of the different power variables of the proposed system are controlled by the load demand, the size of the solar PV system, the size of hydro generator for energy recovery system, the size and state of charge of the battery, and grid electricity price at that ToU period. Therefore, the operation cost will be lessened by the developed optimal energy management scheme.

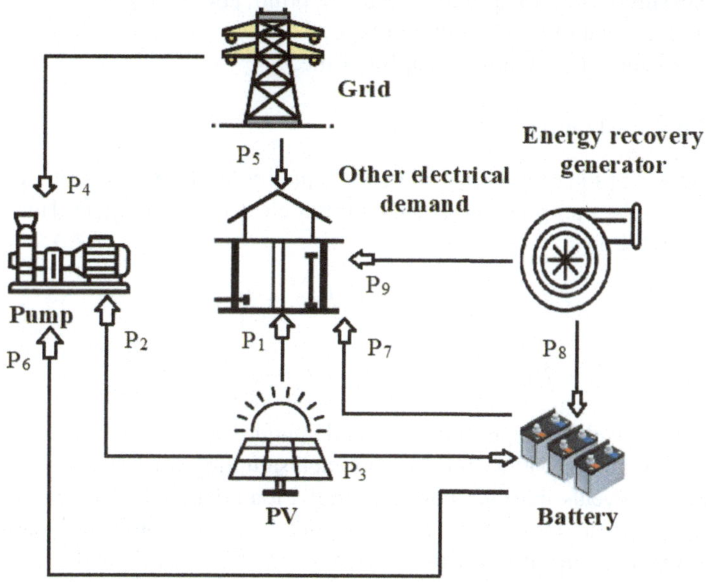

Fig. 1. Proposed grid-connected PV powered marine outfall with energy recovery from wastewater discharge.

The different power flows from Fig. 1 can be defined as follows:

- P_1 is the solar PV power to the marine outfall electrical loads (kW).
- P_2 is the solar PV power to the pumping system demand (kW).
- P_3 is the solar PV power to the batter charging process (kW).
- P_4 is the utility grid power to the pumping system demand (kW).
- P_5 is the utility grid power to the marine outfall electrical loads (kW).
- P_6 is the battery system power to the pumping system demand (kW).
- P_7 is the battery system power to the marine outfall electrical loads (kW).
- P_8 is the energy recovery system power to the battery charging process (kW).
- P_9 is the energy recovery system power to the marine outfall electrical loads (kW).

2.2 Objective Function

The proposed energy management system aims to lessen the grid energy cost, operating under ToU tariff scheme, while maximizing the power from the PV in supplying both the pumping and the marine outfall electrical needs, for the selected simulation horizon. This can then be expressed as:

$$f = \min \sum_{j=1}^{N} [\rho_j (P_{4(j)} + P_{5(j)})] \quad (1)$$

where f is the cost function to be minimised.

- ρ_j is the grid energy cost under the applicable ToU tariff;
- j is the measured sampling range during j^{th} optimization;
- N is the overall quantity of sampling range;
- Δt is the considered simulation sampling time.

2.3 Variable Constraints

Load Balances. For the proposed system, the equality limits are set in order to meet the demands of both the pumping and marine outfall electrical equipment. Hence, this can be expressed as follows:

$$P_{P(j)} = P_{2(j)} + P_{4(j)} + P_{6(j)} \quad (2)$$

$$P_{M(j)} = P_{1(j)} + P_{5(j)} + P_{7(j)} + P_{9(j)} \quad (3)$$

Equation (2) reveals that the pumping's load demand, P_{PJ}, is to be met by a mixture of the solar PV, grid and/or battery powers, for each sampling time "j".

Equation (3) reveals that the marine outfall's load demand, P_{MJ}, can be met by a mixture of the PV, grid, battery and energy recovery system, for each jth sample.

Power Sources Inequality Limits. The following limits can take place per sampling time (j):

- The sum of solar PV power flows for supplying both demands and/or recharging the battery, must not be more than the maximum solar PV power generated in the considered specific sampling range. The maximum solar PV power is controlled by the size of the solar PV system as well as the available solar resources. This can be expressed as:

$$P_{1(j)} + P_{2(j)} + P_{3(j)} \leq P_{PV(j)}^{\max} \quad (4)$$

- The sum of power flows from the grid should be less or equals to the maximum allowable power consumed by the pumping unit and marine outfall's load demand. This can be expressed as:

$$P_{4(j)} + P_{5(j)} \leq P_{Batt(j)}^{Rated} \quad (5)$$

- The sum of power flows from the battery storage unit should be less or equals to the maximum allowable power consumed by the pumping unit and marine outfall's load demand (based on the capacity of the battery). This can be expressed as:

$$P_{6(j)} + P_{7(j)} \leq P^{Rated}_{Batt(j)} \tag{6}$$

- The sum of the power supplied by the energy recovery system (generator) to meet the marine outfall's load demand and/or to recharge the battery must be less or equal to the maximum generated power of the energy recovery system. This can be expressed as:

$$P_{8(j)} + P_{9(j)} \leq P^{Rated}_{Gen} \tag{7}$$

Battery Energy Dynamic. During charging or discharging of the battery, the state of charge (SoC) of the battery can be as follows:

$$SoC_{(j)} = SoC_{(0)} + \frac{\Delta t}{C_n} \times \left(\eta_{Ch} \times \sum_{j=1}^{N} (P_{3(j)} + P_{8(j)}) - \frac{\sum_{j=1}^{N} P_{6(j)} + P_{7(j)}}{\eta_{Disc}} \right) \tag{8}$$

where C_n is the nominal capacity of the battery system (kWh); $SoC_{(0)}$ is the initial state of charge at the beginning of every sampling time; η_{Disc} and η_{Ch} and are the battery's discharging and charging efficiencies, respectively.

Variable Boundaries. All power generation equipment must operate within the allowable limits based on their ratings or sizes. These boundaries are therefore expressed using Eq. (9) for all the power control variable:

$$0 \leq P_{i(j)} \leq P^{\max}_{i(j)} \tag{9}$$

where P_i symbolizes different control variable ranging from P_1 to P_9; $P^{\max}_{i(j)}$ is the maximum or rated power that can be generated by the solar PV modules and hydro-generator; which depends on the available solar and hydro resources. However, the maximum power for both the grid and the battery storage system can be expressed as $P^{Rated}_{i(j)}$ since these control variables are depending on the design capacity instead of available resources.

The battery SoC limits are simulated to be between the initial limit and maximum limit and can be expressed as follows:

$$SoC_0 \leq SoC_j \leq SoC^{\max} \tag{10}$$

Exclusive Power-flow. The power cannot be consumed from the battery and be supplied to the battery at the same time. Hence, the product between the consumed battery power and the supplied battery power is expressed to be zero as revealed by Eq. (11) below:

$$(P_{3(j)} + P_{8(j)}) \times (P_{6(j)} + P_{7(j)}) = 0 \tag{11}$$

Battery's Initial and Final State Condition. For the repeated optimal control algorithm, the initial SoC of the battery should be equal at both the start and the end of the simulation horizon. Hence, this is then expressed as follows:

$$\sum_{j=1}^{N}\left(-P_{3(j)} + P_{6(j)} + P_{7(j)} - P_{8(j)}\right) = 0 \tag{12}$$

Solver Selection. The developed model consists of multiple functions that includes nonlinear functions as well. Hence, "*fmincon*" MATLAB solver has been selected to be determine the optimal solution [18]. This MATLAB solver is expressed as:

$$\min_{x} f(x) \text{ Subject to : } \begin{cases} c(x) \leq 0 \\ c_{eq}(x) = 0 \\ A \cdot x \leq b \\ A_{eq} \cdot x = b_{eq} \\ l_b \leq x \leq u_b \end{cases} \tag{13}$$

where: A and A_{eq} are the inequality and equality matrices, respectively; x, b, b_{eq}, l_b, and u_b are vectors; $c(x)$ and $c_{eq}(x)$ are functions that return vectors and $f(x)$ is a function that returns a scalar.

3 Case Study Description

The existing marine outfall wastewater plant situated in Cape Town (city of South Africa). This has been selected as a case study during simulation. The sections below explain other necessary data as used during simulation.

3.1 Load Profiles

For the proposed case study, the electrical load demand profile and the marine outfall's pumping demand profile, are as shown in Fig. 2.

The electrical load demand profile of the outfall is as a result of both the primary and secondary treatment plants. The primary treatment plant is used to remove debris within the wastewater while the secondary treatment plant is used to remove the suspended solids and organic material, and further disinfects and reduces bacteria within the wastewater and tertiary treatment. This also comprises of the elimination of suspended material using sand filters, as this treatment is a biological process and chemical precipitation is used to eliminate harmful chemical substances [15].

The energy is mostly needed for the demands such as pumping, primary and secondary treatments, space heating, and sludge heating and disposal, continuous ventilation.

3.2 Solar Resources

The solar resource for the specific location, used for sizing the PV system as well as for simulation purpose is available from Ref [19].

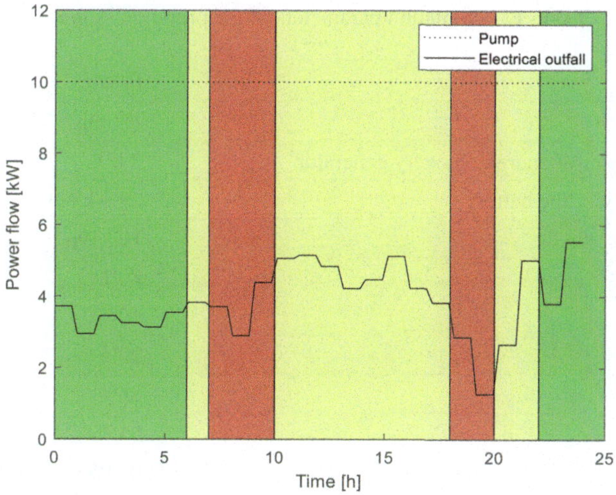

Fig. 2. Load demand profiles (pumping and electrical in marine outfall).

3.3 System Components' Sizes

The approach used for sizing the solar PV modules, battery and hydroelectric generator from the energy recovery system, are described in reference [17, 20] and [21] respectively. The size of different components of the proposed system as well as further simulation parameters are given in Table 1. The ToU structure for the selected day is given in Ref [20].

4 Discussion of the Simulation Results

This section describes the simulated results to reveal the performance of the proposed system when operating under the control of the developed optimal energy management model. The results are discussed in pursuant to the different power flows, as analyzed from the two main demands point of view.

The case whereby the utility grid has been used as the only source of electricity for supplying both the pumping and the marine outfall electrical demand, has been simulated as a baseline for comparison purposes. Hence, the baseline grid cost will then be compared with the results of the proposed system when operating under the developed optimal energy management scheme for supplying the same load demand.

4.1 Baseline: Grid Alone to Supply Both Demands

In this case, the utility grid power is used as a solitary source of power to meet both the pumping and existing electrical demands of the marine outfall, as shown on Fig. 3. Hence, the grid will follow the load profile to reliably meet the overall load demand. Hence, the operation cost resulting from sole grid usage will be compared to the resulting cost of using grid in hybrid with the PV, battery and the energy recovery system.

Table 1. Simulation parameters of the proposed system.

Item	Figure
Sampling time (Δt)	15 min
Rated output power of energy recovery generator	3 kW
Nominal capacity of the battery	29.6 kWh
PV rated power	10 kWp
$SoC_{(0)}$	80%
SoC^{max}	100%
$SoC^{\min}$	20%
η_{Char}	85%
$\eta Disch$	95%
ρ_k	0.20538 $/kWh
ρ_0	0.03558 $/kWh
ρ_s	0.05948 $/kWh

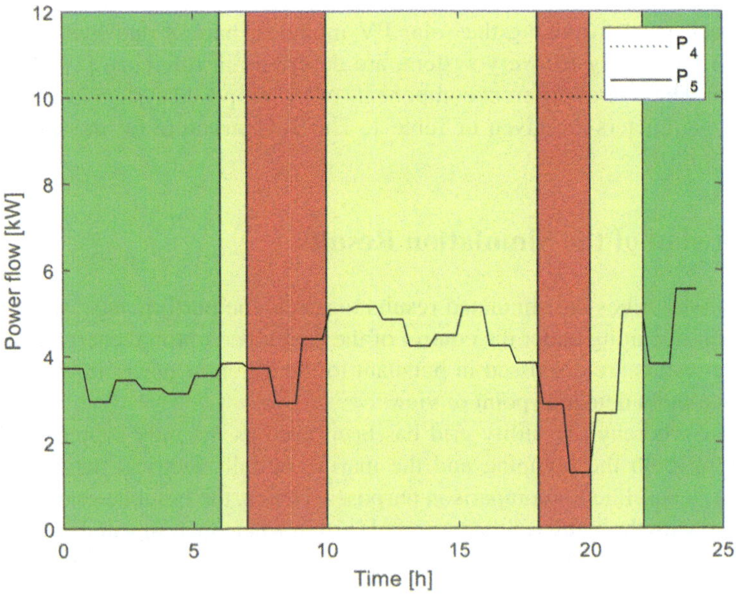

Fig. 3. Load exclusively supplied by the grid.

4.2 Optimal Energy Management of the Proposed Grid Connected System with Energy Recovery Supplying the Pump

Optimal Energy Management as Seen Within the Pumping Unit
Diverse power flows from the solar PV module, the grid and from the battery used to supply the pumping system, are as shown in Fig. 4. The operation will be discussed in each of the different time pricing durations as follows:

- Operation during the First Off-peak Pricing Duration from 00:00 to 06:00

During the first off-peak pricing duration, the pumping load is supplied by the battery and the grid as shown from Fig. 4. The is no power from the PV because of the lack of solar resource. The corresponding decrease in the battery's SoC is shown on Fig. 5.

- Operation during the first standard pricing duration from 06:00 to 07:00

During the first standard pricing duration, the pumping load is supplied by the battery and the grid as shown from Fig. 4. There is still no power from the PV because of the lack of solar resource.

- Operation during the first peak pricing duration from 07:00 to 10:00

During the first peak pricing duration, the pumping load is mainly supplied by the PV, which starts producing power, and the deficit of power is supplied from the grid as shown from Fig. 4. This strategy assists in reducing the operation cost. There is no power supplied from the battery as the SoC is at its lowest as shown on Fig. 5.

- Operation during the second standard pricing duration from 10:00 to 18:00

During the second standard pricing duration, the pumping load is supplied by the PV and the grid as shown from Fig. 4. However, towards the end of this pricing duration, there is a decrease in solar resource. Therefore, a small contribution from the battery is added to the PV and the grid, supporting the pumping energy requirement. It can also be seen from Fig. 5, looking at the battery state of charge, that the battery is being recharged by the PV as well as by the energy recovered from the hydro-generator.

- Operation during the second peak pricing duration from 18:00 to 20:00

During the second peak pricing duration, the pumping load is supplied by the battery and the grid as shown from Fig. 4. The is no power from the PV because of the lack of solar resource. The corresponding decrease in the battery's state of charge is shown on Fig. 5. This is to prepare for the coming second peak pricing duration.

- Operation during the third standard pricing duration from 20:00 to 22:00

During the third standard pricing duration, the pumping load is solely supplied by the grid as shown from Fig. 4. The battery is being recharged to its maximum capacity by the energy recovery system; the corresponding increase in the SoC can be seen from Fig. 5.

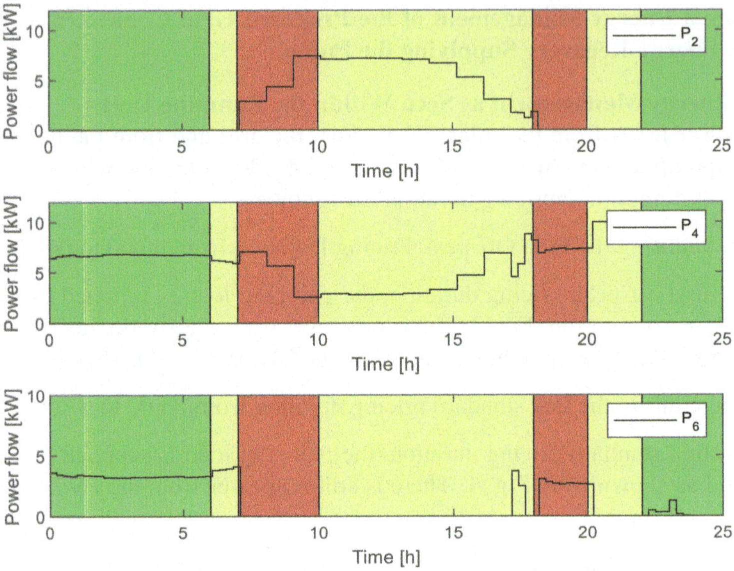

Fig. 4. Power flows as seen from the pumping's side.

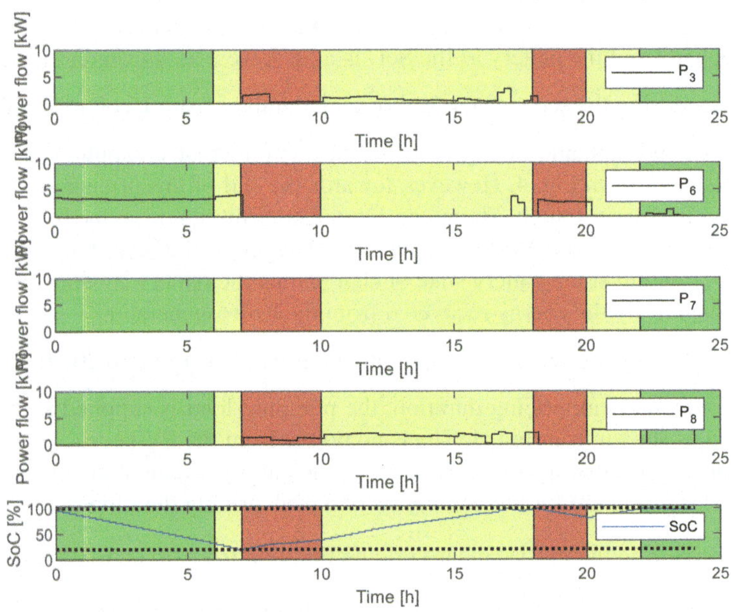

Fig. 5. Battery SoC as well as changing and discharging power flows.

- Operation during the second off-peak pricing duration from 22:00 to 00:00

During the second off-peak pricing duration, the pumping load is supplied with a minimum contribution from the battery and the grid as shown from Fig. 4. The final fixed state of the battery is met as modeled in Eq. 12.

Optimal Energy Management as Seen from the Marine Outfall Electrical Demand
Figure 6 shows the different power flows from the PV, the grid, energy recovery system and from the battery used to supply the pumping system. The operation will be discussed in each of the different time pricing durations as follows:

- Operation during the first off-peak pricing duration from 00:00 to 06:00

During the first off-peak pricing duration, the marine outfall load is supplied by the energy recovery system and the grid as shown from Fig. 4. The is no power from the PV because of the lack of solar resource and no contribution from the battery.

- Operation during the first standard pricing duration from 06:00 to 07:00

During the first standard pricing duration, the marine outfall load is supplied by the energy recovery system and the grid as shown from Fig. 6. There is still no power from the PV because of the lack of solar resource.

- Operation during the first peak pricing duration from 07:00 to 10:00

During the first peak pricing duration, the marine outfall load is mainly supplied by the PV and the energy recovery system, and the deficit of power is supplied from the grid as shown from Fig. 6. This strategy assists to minimize the operation cost which is high.

- Operation during the second standard pricing duration from 10:00 to 18:00

During this second standard pricing duration, the marine outfall load is mainly supplied by the PV and the energy recovery system, and the deficit of power is supplied from the grid as shown from Fig. 6. The contribution from the grid is higher as compared to the previous pricing duration.

- Operation during the second peak pricing duration from 18:00 to 20:00

During the second peak pricing duration, the marine outfall load is mainly supplied by the energy recovery system operating in conjunction with the grid, which is minimized as shown from Fig. 6. The is no power from the PV because of the lack of solar resource.

- Operation during the third standard pricing duration from 20:00 to 22:00

During the third standard pricing duration, the marine outfall load is mainly supplied by the energy recovery system operating in conjunction with the grid, which is minimized as shown from Fig. 6.

- Operation during the second off peak pricing duration from 22:00 to 00:00

During the second off peak pricing duration, the same operation strategy is applied as for the previous pricing duration.

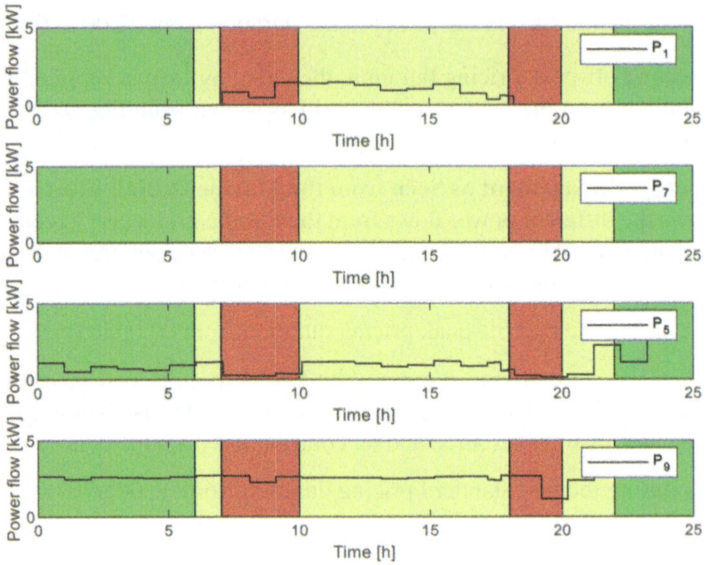

Fig. 6. Power flows as seen from the marine outfall electrical demand's side.

4.3 Daily Economic Analysis

Table 2 summarizes the daily operation cost comparison between the grid and the proposed hybrid system with energy recovery generator. The cost achieved by the hybrid system without energy recovery is also considered. From the results, substantial cost saving can be achieved using the proposed system operating under the developed optimal energy management model.

Table 2. Costs comparison.

Supply option	Daily operation cost (US$)	Saving
Baseline	48.58	-
Hybrid system without energy recovery	15.70	68.68%
Hybrid system with energy recovery	11.95	75.5%

5 Conclusion and Recommendations

This paper focused on the development of optimal energy management model to lessen the operation cost of the existing marine outfall as supplied by the combination of the grid-connected solar PV-battery based system and wastewater energy recovery system. The model has been applied to a case in South Africa; and the simulation results were investigated and interpreted in comparison to the current case whereby the utility grid power is solely used to meet the overall demand of the marine outfall's system. The simulation results showed that the application of the developed optimization model on the proposed system, leads to a daily energy cost saving as achievable under the marine outfall's operation as compared to the case where the utility grid is solely used to supply similar load demand.

The simulation results revealed that the energy cost saving of 75.5% is achievable through the use of the proposed system when operating under the control of the developed optimal energy management model, as compared to the grid used as baseline.

In this work, the techno-economic analysis was based mainly on analysis of the operational cost saving accrued per day. It should be mentioned that the financial performance of the proposed system is also dependant on other parameters such as the initial cost of different components. It is also dependant on time value of the investment as well as operation and maintenance costs. Therefore, for future work, the lifecycle cost assessment ought to be performed to better analyse the financial performance of the proposed system.

References

1. Kusakana, K., Vermaak, H.J.: Feasibility study of hydrokinetic power for energy access in rural South Africa. In Proceedings of the IASTED Asian conference, power and energy systems, pp. 433–438 (2012)
2. Paish, O.: Small hydro power: Technology and current status. Renew. Sustain. Energy Rev. **6**(6), 537–556 (2002)
3. Berga, L.: The role of hydropower in climate change mitigation and adaptation: A review. Engineering **2**(3), 313–318 (2016)
4. Kusakana, K.: Optimization of the daily operation of a hydrokinetic–diesel hybrid system with pumped hydro storage. Energy Convers. Manage. **106**, 901–910 (2015)
5. Kusakana, K.: Optimal scheduling for distributed hybrid system with pumped hydro storage. Energy Convers. Manage. **111**, 253–260 (2016)
6. Kusakana, K., Vermaak, H.J.: Hydrokinetic power generation for rural electricity supply: Case of South Africa. Renew. Energy **55**, 467–473 (2013)
7. Thorburn, K.: Electric energy conversion systems: wave energy and hydropower. Acta Univ. Upsaliensis, PhD diss. (2006)
8. Shiono, M., Katsuyuki S., Seiji K.: An experimental study of the characteristics of a Darrieus turbine for tidal power generation. Electr. Eng. Japan **132**(3), 38–47 (2000)
9. Griffin, F.M.: Feasibility of energy recovery from a wastewater treatment scheme. Proc. Inst. Mech. Eng. A J. Power Energy **214**(1), 41–51 (2000)
10. Mbele, L., Kusakana, K.: Overview of conduit hydropower in South Africa: Status and applications. In: 2017 IEEE PES PowerAfrica, pp. 385–390. IEEE (2017)
11. Mbele, L.N., Kusakana, K., Koko, S.P.: Simulations and experimental validation of Pico conduit pressure hydropower systems with battery storage. J. Energy Storage **26**, 100976 (2019)
12. Mbele, L., Kusakana, K.: Modelling water energy recovery system using MATLAB/Simulink software. Adv. Sci. Lett. **24**(11), 8209–8214 (2018)
13. Mbele, L.N., Kusakana, K.: Model-based design of a conduit Pico hydropower system. In: 2018 IEEE PES/IAS PowerAfrica, pp. 769–774. IEEE (2018)
14. Mbele, L.N., Kusakana, K., Koko, S.P.: Experimental analysis of small conduit pressure hydropower systems. In 2019 International Conference on the Domestic Use of Energy (DUE), pp. 16–19. IEEE (2019)

15. Moselane, P., Kusakana, K., Koko, S.P.: Using marine outfall wastewater discharge systems for electricity generation. In: 2019 Open Innovations (OI), pp. 206–210. IEEE (2019)
16. Moselane, P., Kusakana, K., Koko, S.P.: Techno-economic analysis of marine outfalls wastewater discharge for electricity generation. In: 2022 IEEE PES/IAS PowerAfrica, pp. 1–5. IEEE (2022)
17. Moselane, P., Koko, S.P., Kusakana, K.: Modelling and simulation of a hydropower system using wastewater discharged from Marine outfalls. Energy Rep. **8**, 780–787 (2022)
18. Kusakana, K.: Energy management of a grid-connected hydrokinetic system under time of use tariff. Renew. Energy **101**, 1325–1333 (2017)
19. Hohne, P.A., Kusakana, K., Numbi, B.P.: Optimal energy management and economic analysis of a grid-connected hybrid solar water heating system: A case of Bloemfontein, South Africa. Sustain. Energy Technol. Assess. **31**, 273–291 (2019)
20. Numbi, B.P., Malinga, S.J.: Optimal energy cost and economic analysis of a residential grid-interactive solar PV system-case of eThekwini municipality in South Africa. Appl. Energy **186**, 28–45 (2017)
21. Kusakana, K.: Optimal energy management of a grid-connected dual-tracking photovoltaic system with battery storage: Case of a microbrewery under demand response. Energy **212**, 118782 (2020)

Reduction of Exhaust Emission of Agricultural Diesel Engine Refuelling with Coconut Oil Biodiesel and Performance

Karoon Fangsuwannarak[1](✉) and Thipwan Fangsuwannarak[2]

[1] School of Mechanical Engineering, Institute of Engineering, Suranaree University of Technology, Nakhon Ratchasima, Thailand
karoon@sut.ac.th

[2] School of Electrical Engineering, Institute of Engineering, Suranaree University of Technology, Nakhon Ratchasima, Thailand

Abstract. This study presents the production of coconut oil biodiesel (COB) by using the trans-esterification process. The combustion efficiency and exhaust gas emission of an agricultural diesel engine using COB compared with diesel fuel were investigated. The 4-stroke cycle, single-cylinder diesel engine with naturally aspirated air and fuel injection degree 24 bTDC has been tested at a compression ratio of 17. The engine test was evaluated under the speed at 1,500 rpm with various load tests at 3, 6, 9 and 12 kg. The %yield of COB from the production process is found to be 90.52%. The fatty acid content is 75.85% in COB using Gas Chromatography. COB consists of specific gravity, viscosity, cetane index, and heat value of the fuel were measured according to ASTM standards. A cetane index of COB provides a lower value than that of diesel fuel. It leads to the engine using COB obtaining higher specific fuel consumption value than diesel fuel by 19.67%. However, it is noticed that COB provides outstanding physical properties in lower viscosity than diesel fuel with its containing 47.52% Lauric acid fatty acids, Myristic acid 18.37%, and Palmitic acid 9.24%. COB components with a carbon chain shorter than other fuels result in the lowest fuel calorific value. The combustion engine performance using COB shows a shorter ignition time delay than that using diesel fuel. The COB properties have a direct influence on the reduction of emissions and smoke from the agricultural diesel engine compared with diesel fuel used.

Keywords: coconut biofuel · combustion characteristics · agricultural engine performance

1 Introduction

The largest increase in the number of automobiles for transportation and agriculture machines has resulted in greater demand for petroleum products with the trend of fast depletion of world petroleum reserves [1]. The main point of the world energy demand in the last two decades has encouraged researchers to search for new alternative energy

J. Baeyens et al. (Eds.): ESRE 2023, pp. 113–124, 2025.
https://doi.org/10.1007/978-981-96-4345-5_9

sources such as biofuels, solar energy, wind energy, and biomass [2]. Biodiesel is an alternative fuel due to its various advantages, such as abundant biomass resources, biodegradability, and environmental friendliness over fossil fuel [3]. Moreover, biodiesel predominantly produced from biomass is becoming cost-competitive with fossil fuels due to the widespread availability of biomass resources [4]. At present, coconut oil is a trade commodity to cause direct competition between using coconut oil for biodiesel production and using it for food. This reason leads to an increase in the price of both uses. Nevertheless, raw coconut has a massive energy plantation in Thailand and environmental and local benefits. For the coconut oil biodiesel (COB) production, trans-esterification of coconut oil with methanol to obtain COB was carried out using a potassium hydroxide (KOH) catalyst [5]. ASTM standard fuel tests performed on the COB, in particular. A total of three fuel samples, such as DF (100% diesel fuel), CB5 (5% coconut biodiesel and 95% DF), and CB15 (15% CB and 85% DF). As a result of investigations, there has been a decrease in torque and brake power, while an increase in specific fuel consumption has been observed for biodiesel blended fuels over the entire speed range compared to net diesel fuel. In the case of engine exhaust gas emissions, lower HC, CO, and, higher CO_2 and NO_x emissions have been found for biodiesel blended fuels compared to diesel fuel [6]The effects of parameters related to these processes was studied and optimized, including methanol-to-oil ratio, catalyst concentration, reaction temperature, and reaction time. Methyl ester content of the coconut biodiesel was determined by GC to be 90.28% under the optimum condition. The viscosity of coconut biodiesel products was very close to that of Thai petroleum diesel and other measured properties met the Thai biodiesel (B100) specification [7].

In this study, the coconut oil biodiesel is produced from coconut oils with a compression ratio of 17. The engine test was evaluated under the speed at 1,500 rpm with various load tests at 3, 6, 9, and 12 kg.

2 Materials and Methods

Coconut oil usually exhibits high viscosity values, making them unsuitable to be used as fuel for continuous running. Therefore, a process called trans-esterification is performed so as to reduce the viscosity by treating the coconut oil with alcohol (commonly ethanol or methanol). When a byproduct is obtained in the form of glycerol that needs to be separated. The transesterification reaction is a stage of converting oil or fat into methyl esters.

2.1 Biodiesel Production from Coconut Oil

Coconut oils primarily contain triglycerides, and their chemical structure significantly differs from mineral diesel. Transesterification is an efficient method to convert high viscosity coconut oil into a fuel with chemical properties similar to those of mineral diesel. COB was prepared from coconut oil and methanol by using potassium hydroxide (KOH) as a catalyst through a trans-esterification process. In such biodiesel production, the effect of oil to methanol molar ratio 1:6, catalyst concentration 1% by weight, reaction temperature 60 °C and reaction time 60 min on the yield of transesterification was studied.

The biodiesel yield percentage is the factor that indicates the amount of actual biodiesel yield (Y_{act}) compared with the theoretical biodiesel yield. The theoretical biodiesel yield (Y_{th}) and biodiesel yield percentage (%Y) can be calculated from the following equations.

$$Y_{th} = \frac{N \times MW_{COB} \times m_{Coconut\ Oil}}{MW_{Coconut\ Oil}} \tag{1}$$

$$\%Y = \frac{Y_{act}}{Y_{th}} \tag{2}$$

when,

Y_{act} = actual biodiesel yield (g)
Y_{th} = theoretical yield of biodiesel (g)
N = mole number of biodiesel per mole amount of crude vegetable oil
$m_{Coconut\ Oil}$ = crude palm oil mass (g).
MW_{COB} = molecular weight of coconut oil methyl ester (226.90 g/mol)
$MW_{Coconut\ Oil}$ = molecular weight of crude coconut oil (661.67g/mol)

Figure 1 shows the gas chromatograph profiles of biodiesel, which can be detailed in Table 1. Table 1 shows the main components are palmitic acid (9.24%), Lauric acid (47.52%), and Myristic acid (18.37%). It is noticed that the total FAME of 75.85% experimentally measured in this work is less than that of the standards set of 96.5% by the Department of Energy Business, Thailand. The average molecular weight of coconut methyl ester from Table 1 equal to 226.90 g/mol and the result of the transesterification reaction with coconut oil 5000 g reaction results in actual biodiesel yield (Y act) of 4656.19 g. Thus, the biodiesel yield percentage (%Y) of 90.52% was obtained when the theoretical yield of biodiesel (Y_{th}) was calculated of 5143.80 g.

2.2 Experimental Test Setup and Procedure

In a single-cylinder direct injection diesel engine with a naturally aspirated, water-cooled system was used for the engine test to evaluate engine performance and exhaust emissions. The engine was operated by using diesel fuel and the results were used as reference values. Thus, the results obtained from the engine fuelled with COB were compared with diesel fuel. The engine specification is shown in Table 2. The experimental installation diagram is drawn schematically in Fig. 2. An eddy current dynamometer equipped with a load cell was used to load the engine. The engine was tested at a constant speed of 1500 rpm with varying three-engine operating loads (3, 6, 9, and 12 kg or brake powers of 0.86, 1.71 and 2.57, 3.42 kW), which represents the range of the engine operating condition tested at a compression ratio of 17. Exhaust emissions were measured by standard exhaust gas analysers: Testo 350 for CO and NOx, Horiba Mexa-584L for HC and Wager Digital Smoke Meter Model 6500 for smoke emissions. In each test, the engine was stabilized by the observation of exhaust temperatures and exhaust emissions.

Table 1. Total fatty acid methyl esters (FAME) quantification in COB gas chromatograph together with all the identified compounds with their area and retention time.

Peak NO	Retention Time	Name	C No.: Dbl.Bnd	Molecular Weight	FAME %
1	12.14	Butyric acid	C4:0	102.13	0.05
2	14.64	Caproic acid	C6:0	130.18	0.49
3	16.87	Caprylic acid	C8:0	158.24	7.59
4	18.94	Capric acid	C10:0	186.29	6.14
5	20.005	Undecanoic acid	C11:0	200.324	0.02
6	21.159	Lauric acid	C12:0	214.351	47.52
7	22.39	Tridecanoic acid	C13:0	228.37	0.03
8	23.90	Myristic acid	C14:0	242.40	18.37
9	27.98	Palmitic acid	C16:0	270.45	9.24
10	34.56	Stearic acid	C18:0	298.51	2.75
11	37.34	Oleic acid	C18:1n9c	296.51	6.56
12	41.79	Linoleic acid	C18:2n6c	294.51	1.21

Table 2. Technical specifications of diesel engine test.

Engine	Specification
Make	Kirloskar Oil engines
Engine type	1 cylinder, 4 strokes, Direct injection
Bore	87.5 mm
Stroke	110 mm
Connecting rod length	234 mm
Swept volume	661 cc
Rate output	3.5 kW at 1500 rpm
Compression ratio	12–18
Dynamometer	Eddy current, Water cooled

3 Results and Discussion

3.1 Comparative Physical Properties of Tested Fuels

This research measures the COB blend and diesel fuel. Thailand has set the standard for biodiesel by the Department of Energy Business, Ministry of Energy. This is due to the requirement for quality control for engine use. The experimental results of fuel properties were measured as detailed in Table 3. Kinematic viscosity, specific gravity,

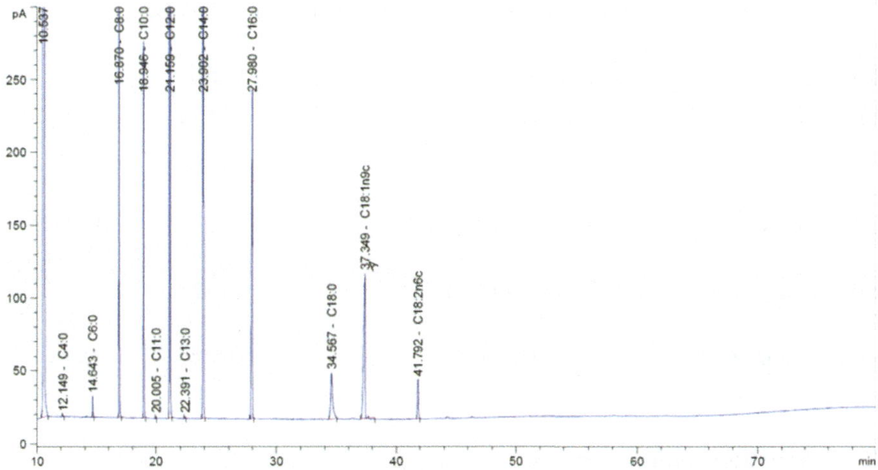

Fig. 1. Gas chromatograph profiles of coconut oil biodiesel (COB).

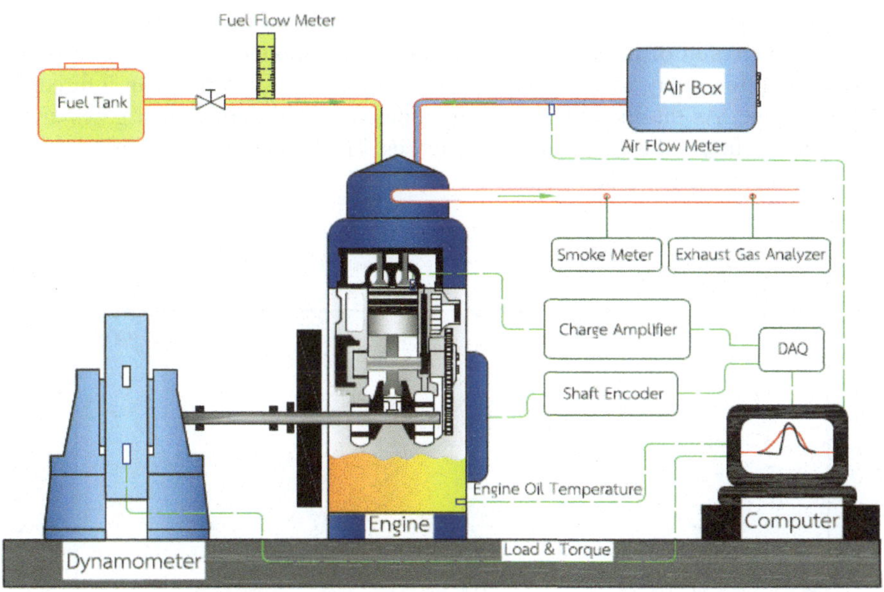

Fig. 2. Diagram of the installation system of the engine test.

flash point, cetane index, and calorific value are obtained in the standard of the criteria set by the American Society of Testing and Materials (ASTM) standard. Each measurement was replicated 3 times to average the data used for the analysis. This viscosity result of COB has related to the higher flash point and its density.

Table 3. Properties of Diesel and Coconut oil biodiesel (COB).

Fuel/Property	Standard	Diesel	COB
Specific Gravity at 15.6 °C	ASTM D1298 0.81 ≤ 0.87	0.830	0.867
Kinematic Viscosity (cSt)	ASTM D445 1.8 ≤ 4.1	3.92	3.62
Flash Point (°c)	ASTM D93 > 52	59.33	104.66
Cetane Index	ASTM D976 > 50	56.82	44.57
Calorific Value (MJ/Kg)	ASTM D240	46.49	38.61

3.2 Combustion Characteristics

Figure 3 shows the variation of the pressure in the cylinder with the degree of the crankshaft. The experimental results show that when the load is increased, the pressure in the cylinder increases and the ignition delay is reduced. The shortest ignition delay is probably the result of increased fuel injection. The amount of fuel injection causes a reduced ignition delay as the load increases are linearly correlated in a direct fuel injection engine. As the load increases, the temperature of the residual gases and the temperature of the combustion chamber increases. The higher the air-gas mixture at the injection stroke thereby reducing the lag.

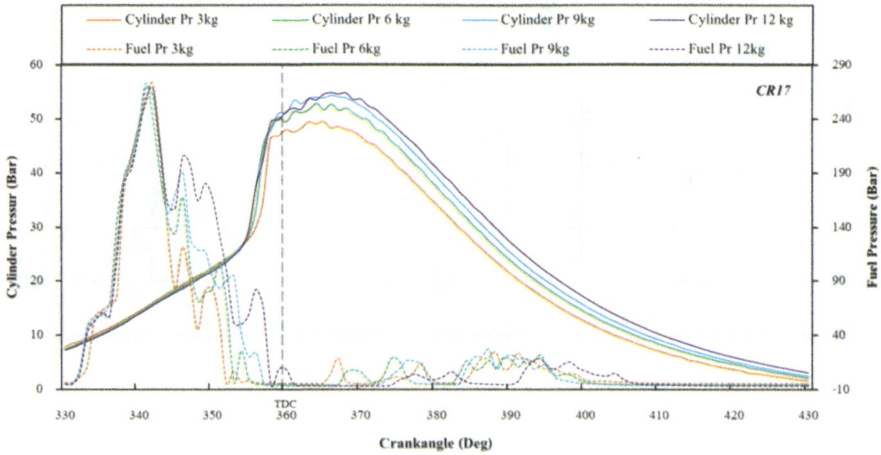

Fig. 3. Variation of the pressure in the cylinder with the degree of the f crankshaft.

The heat release rate and fuel pressure at each degree of a crankshaft of diesel fuel compared with the active load of 3, 6, 9, and 12 kg at compression ratio 17, as illustrated in Fig. 4. It is noted that when the load is increased, the heat release rate is lower in the premix combustion phase and there is a controlled combustion phase by the mixture control) increased instead. The shorter ignition delay is the result of increased fuel injection, which is linearly correlated in direct fuel injection engines. When the load increases, the temperature of the residual gases and the temperature of the combustion chamber increase, causing a higher temperature of the air-gas mixture at the time of injection.

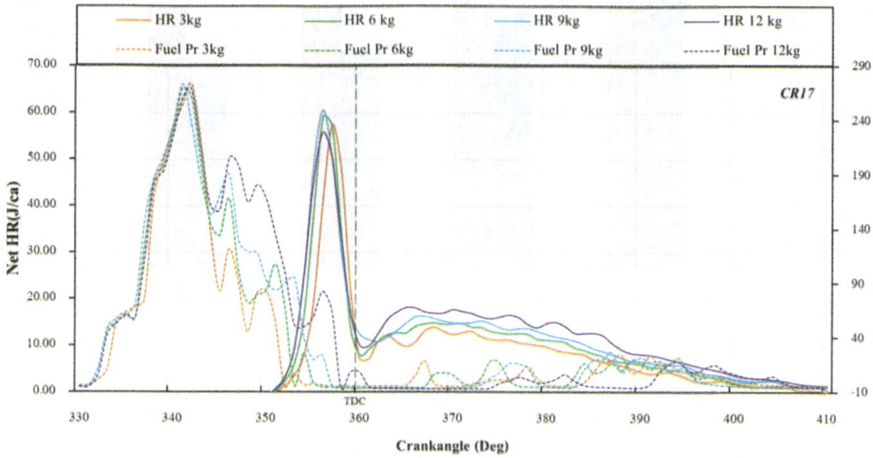

Fig. 4. Variation heat release rate with degree of crankshaft.

3.3 Engine Performance

The results of engine performance were tested, including the combustion feature and pollution caused by incomplete combustion. The comparative results of using diesel with COB in a diesel engine, are shown in the following test results.

Brake-Specific Consumption. Brake Specific Fuel Consumption (BSFC) is the rate of fuel consumption per unit of power generated by the engine within a unit of time. Figure 5 shows BSFC with COB tested has a higher fuel consumption ratio than diesel in all load conditions because the heating value of biodiesel is less than that of diesel fuel. Thus, to obtain the same thermal energy, an engine powered by COB needs a greater quantity of fuel than diesel. BSFC with tested COB becomes closer and approaches BSFC with tested diesel fuel. It is due to the increase in the ratio resulting in the shorted ignition delay, which can compensate for the disadvantages viewpoints of the lower COB viscosity and the lower cetane index than that of diesel as shown in Table 3.

Brake Thermal Efficiency. The Brake Thermal Efficiency (BTE) of the engine is the ability to change the heat energy obtained from the fuel into mechanical energy. As shown in Fig. 6. The BTE value of the engine-fuelled COB is similar to that of the diesel at the compression ratio of 17. However, the engine fuelled by COB obtained a lower BTE value than diesel owing to the poor performance of COB spray at a lower temperature in the combustion chamber. This condition results in a lower cetane value of using COB than diesel due to the presence of inconsistent combustion.

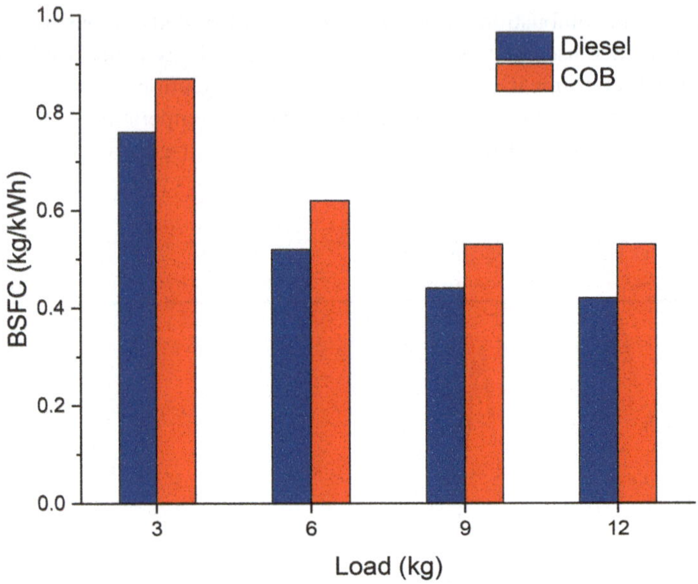

Fig. 5. Variation Brake Specific Fuel Consumption with the load.

3.4 Emission Analysis

Oxide of Nitrogen Emissions (NO_x). Nitrogen oxide (NO_x) gas is the main exhaust component from engine combustion consisting of nitric oxide (NO) and nitrogen dioxide (NO_2). High combustion temperature can affect the rise of NO_x emission due to the high possibility of nitrogen bonding with oxygen. The oxygenated nature of COB provides a slightly higher adiabatic flame temperature resulting in better combustion complete than the diesel test. Figure 7 shows NO_X emission with varying loads and combustion ratios for COB and diesel tests. It is noted that under all engine load tests. Thus, the low viscosity of COB and high flash point can cause the decay of fuel injection and subsequently affect a decreased in NO_X emission.

Carbon Monoxide Emissions (CO). Carbon monoxide emissions are mostly dependent on the fuel properties, the amount of oxygen, the combination of air and fuel, temperature and turbulent airflow in the combustion chamber. CO emissions with varying loads and combustion ratios for COB and diesel tests is shown in Fig. 8. Comparing the CO emissions of biodiesel and diesel fuel, it is found the decrease of CO emission from COB test comparing with diesel test and engine load conditions. It is mostly possible due to C-O bonds lightly occurring at combustion temperatures and its shorter ignition delay also results in better the complete fuel combustion.

Hydrocarbon Emissions (HC). Micro-particles of hydrocarbons are an organic compound that can lead to incomplete combustion of the fuel. Figure 9 shows HC emissions from the engine fuelled by COB and diesel fuel at various loads. The results of decreased HC micro-particles significantly indicate the fuel combustion efficiency due to its oxygenated nature of COB leading to the effectively complete combustion. It is consistent with the availability of oxygen atom in biodiesel.

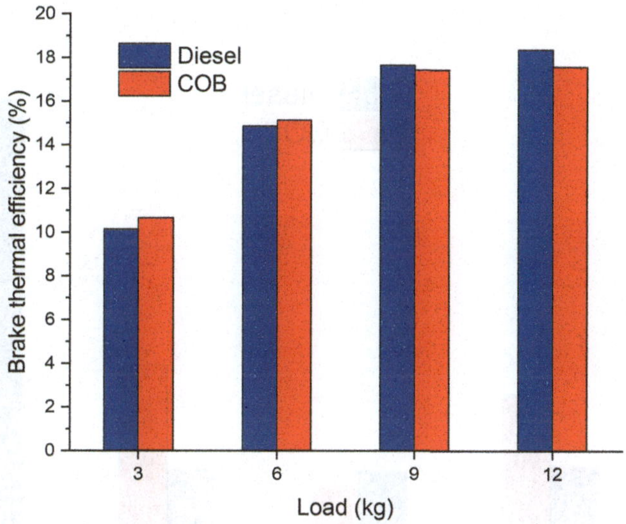

Fig. 6. Variation brake thermal efficiency with the loads.

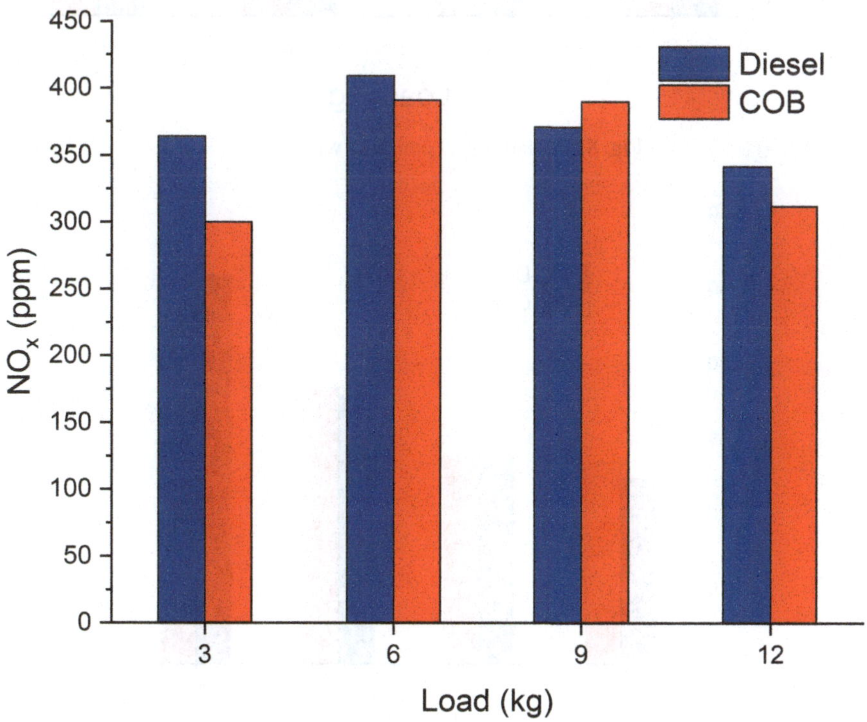

Fig. 7. Variation NOx emission with the loads.

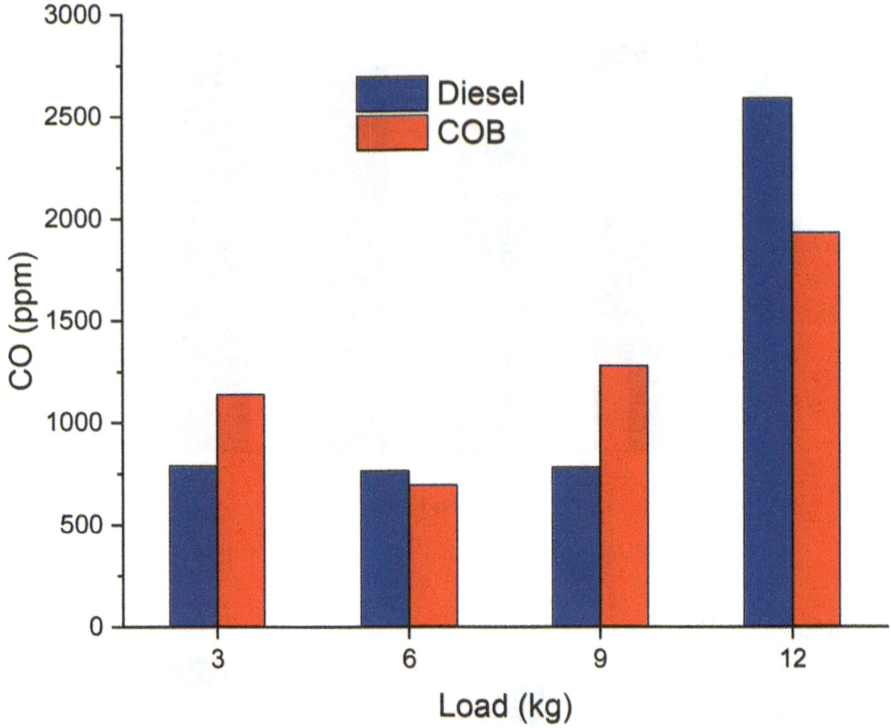

Fig. 8. Variation CO emission with loads.

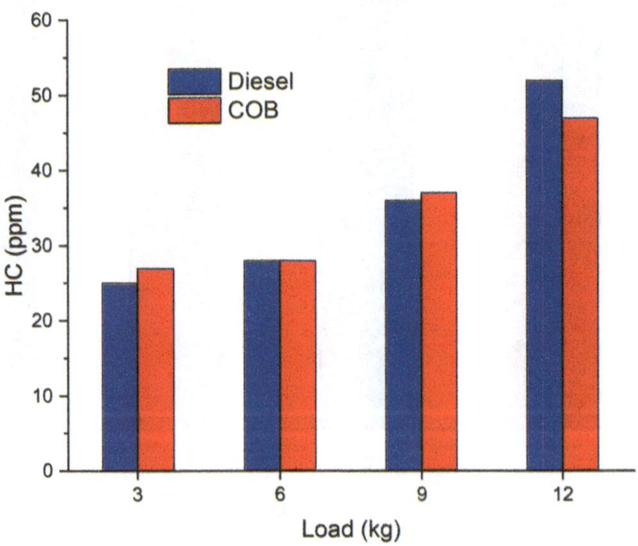

Fig. 9. Variation HC emission with loads.

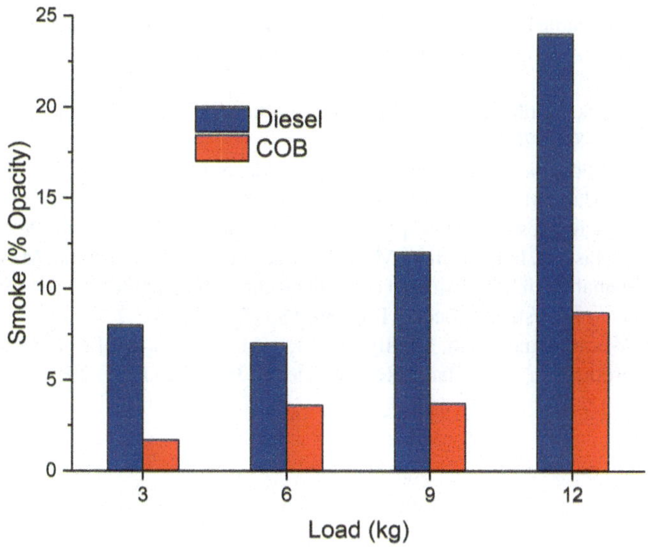

Fig. 10. Variation of smoke opacity with loads.

Smoke emissions. The main cause of smoke formation is the incomplete combustion of a hydrocarbon in fossil fuels due to the chemical reaction of carbon components. Smoke emissions are directly related to the results of CO and HC emissions. The COB test results indicate that smoke emissions are lower than the diesel test. The smoke emission usually increases with increasing load tests because of the rich fuel mixture as shown in Fig. 10.

4 Conclusions

The engine emission characteristics were verified under the engine tests, which were fuelled by COB and diesel. In the case of COB used, it was found the reduction of kinematic viscosity by 30.13%, whereas the flash point was increased by 43.31% compared with the test using diesel fuel. Meanwhile, the brake-specific fuel consumption increased by 22.70%, whereas brake thermal efficiency increased by 0.37%, compared with the diesel test. When using COB found that the NO_X, HC, CO, and smoke emissions were reduced by 4.08, 1.38, 2.57, and 188.14%, respectively, compared with using diesel fuel.

Acknowledgements. This study was supported by Suranaree University of Technology.

References

1. Karthikeyan, R., Mahalakshmi, N.V.: Performance and emission characteristics of a turpentine diesel duel fuel engine. Energy **32**, 1202–1209 (2005)

2. Panwar, N.L., Shrirame Hemant, Y., Rathore, N.S., Sudhakar, J., Kurchania, A.K.: Performance evaluation of a diesel engine fueled with methyl ester of castor seed oil. Apply Thermal Eng. **30**, 245–249 (2010)
3. Nagarajan, Mahua (Madhuca indica): Seed oil: a source of renewable energy in India. J. Sci. Ind. Res. **64**, 890–896 (2005)
4. Demirbas, A.: Global biofuel strategies. Energy Educ. Sci. Technol. **17**, 27–63 (2008)
5. Alkabbashi, N., Alam, M.Z., Mirghani, M.E.S., Al-Fusaiel, A.M.A.: Biodiesel production from crude palm oil by transesterification process. J. Appl. Sci. **9**, 3166–3170 (2009)
6. Liaquat, A.M., Masjuki, H.H., Kalam, M.A., Rizwanul Fattah, I.M., Hazrat, M.A., Varman, M., Mofijur, M., Shahabuddin, M.: Effect of coconut biodiesel blended fuels on engine performance and emission characteristics. Procedia Eng. **56**, 583–590 (2013)
7. Nakpong, P., Wootthikanokkhan, S.: High free fatty acid coconut oil as a potential feedstock for biodiesel production in Thailand. Renew. Energy **35**, 1682–1687 (2010)

An Inverted Solar Distillation Still; with Integrated Heating Technique Applicable in Cold Climates

Niladri Sarkar[1], Aneesh Somwanshi[2], and Anil Kr. Tiwari[1(✉)]

[1] Department of Mechanical Engineering, National Institute of Technology, Raipur, India
aktiwari.mech@nitrr.ac.in

[2] Department of Mechanical Engineering, MATS University, Raipur, India

Abstract. In present work authors presented an inverted absorber solar still which provides distilled water and simultaneously utilized for space heating of a representative room of size 3.05 m × 3.05 m × 3.05 m. Evaporation of water takes place by solar energy and that condenses under copper metallic condensing cover that is releasing heat to the room for space heating. The potable water finally is drained out to be used, vapor is getting condensed by releasing its latent heat to cover that finally releases the heat to the room as space heating particularly in winter months. Thus, generation of potable water and space heating is done simultaneously. The mathematical model of the proposed device has been developed. Numerical computations have been performed to determine annual amount of distillate output produced and the amount of heat given to room during winters for the climate of Delhi, India. It is seen that more water depth in still basin reduces the total distillate output from still, but due to more heat stored in water at higher depths the rate of heat released for space heating during off sunshine hours is more. The total annual distillate output produced by still in Delhi, India is 1215 L/m^2 and the total heat energy released into space is 305.4 kWh. This way about 200 kg of coal and 484 kg of CO_2 has been prevented from burning and liberated in to atmosphere respectively for a space size of 3.05 m × 3.05 m × 3.05 m.

Keywords: Space heating · Solar distillation · Inverted absorber

1 Introduction

Solar distillation method is an easy, small scale and cost-effective technique of providing safe water at homes or in small scale communities. Distillation is one of many processes that can be used for water purification. The process requires energy input as heat and the solar radiation can be the source of energy. In this process, water is evaporated, thus separating water vapor from dissolved matter. The vapors get condensed as pure water. The use of solar distillation system has some difficulties which sometimes limits the use of this technique for large scale production. The low production output and weather-related problems are some of the major disadvantages with solar distillation process. However, in spite of this major disadvantage still due to the use of inexpensive solar energy the process is found to be suitable for the places receiving adequate amount of solar radiation.

J. Baeyens et al. (Eds.): ESRE 2023, pp. 125–136, 2025.
https://doi.org/10.1007/978-981-96-4345-5_10

The main governing factor responsible for getting yield from solar still is the temperature difference between the basin water and glass cover. The more temperature difference enhances the production output from solar still. The researchers all over the world are continuously trying to increase this temperature difference. This difference in temperature can be increased either by passive or active ways. In active solar stills an extra -thermal energy is fed into the basin of still by external mode. The glass cover cooling method is also one of the active methods of increasing the temperature difference. Somwanshi et al. [1] integrated single slope solar still with desert cooler, the cooled water is allowed to flow over cover of still. They found about 58% increase in distillate output of solar still. Tiwari et al. [2] suggested to integrate the garden fountain with small solar distillation plant. They found the method to be employed in small scale. The coolness of water stored in reservoir of fountain can be used to enhance the output of plant. Passive solar distillation includes the conventional single slope and double slope solar stills. Some other design modifications include spherical shape solar still [3], hemispherical shaped solar still [4], triangular pyramid shaped solar still [5], V shaped solar still [6], conical shaped solar still [7], wick type solar still [8], stepped basin solar still [9] The inverted absorber solar still [10, 11] is another modification in the design of solar still. In inverted absorber solar still is a concentrating type of solar still which consists of a conventional curved reflector under metallic basin. Unlike the conventional solar still the inverted absorber solar still has its different surface receiving the solar radiation from the surface that is used for condensation. The solar radiation is received from the bottom of the absorbing plate and condensation takes place on inner surface of the metallic condensing cover made by copper sheet (Fig. 1). This type of solar still has the advantage of high absorptivity and low bottom loss coefficient of absorbing plate in comparison to conventional solar still.

In inverted absorber solar still the evaporated vapor condenses in the inner surface of the metallic plate releasing its latent heat of vaporization. The condensate trickles down and finally collected through drainage pipe provided at the lower end. The latent heat of vaporization in the metallic plate goes wastage into the atmosphere. To utilize this latent heat of vaporization Tiwari et al. [12] designed multiple inverted solar still in which the latent heat released to the condensing cover of first basin is utilized in the heating water in second basin and from second the heat is utilized in third basin.

In present work the authors suggested this heat to be utilized for space heating during winters. The still generates pure water throughout the year additionally it can be utilized to heat room during winters. The inverted still can be installed in floor of the room as shown in Fig. 1. The reflector part kept outside at an angle equal to latitude of the place to receive maximum solar radiation throughout the year. The reflected solar radiation absorbed by the absorber plate results in heating water placed in the basin of solar still. The evaporated vapor gets condensed and the latent heat of vaporization goes into the condenser thereafter the heat from cover transfers into the room. The size (width) of the solar still depends on the required heating load as well as the quantity of the pure water required. However, while deciding the size of still, one has to look after the area available inside room. Authors suggested two different strategies for winters and summers during winters the still generates distillate as well as the heated condenser plate transfers heat to the room. During other months of a year when the space heating is not required the heat from condenser is released into the atmosphere through window through a temporary ducting arrangement provided in the room. In present work authors presented the mathematical model of the inverted solar still. Performed numerical computations to determine the daily, monthly and annual output of solar still & the daily, monthly and annual rate of heat energy transferred to room.

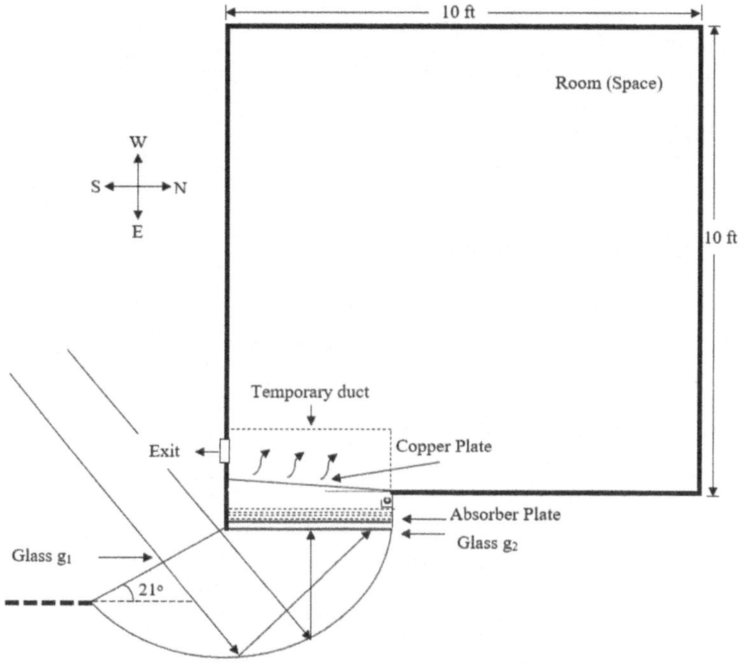

Fig. 1. Schematic diagram of the proposed setup

2 Mathematical Analysis

The following assumptions have been made in order to write the energy balance equations.

a) The heat capacity of glass cover, absorbing material and condensing cover and insulation is negligible.
b) The solar still is vapor proof
c) The aperture area is same as that of absorber plate and concentration ratio is 1:1
d) The temperature of room is considered equal to ambient temperature

Energy balance of glass cover g2

$$h_{rpg2}(T_p - T_{g2}) = U_r(T_{g2} - T_a) \tag{1}$$

Energy balance of absorber plate

$$\tau_{g1}\tau_{g2}\alpha_p r^N I = h_{cpw}(T_p - T_w) + h_{rpg2}(T_p - T_{g2}) \quad (2)$$

Energy balance of water mass

$$M_w c_w \frac{dT_w}{dt} = h_{cpw}(T_p - T_w) - h_1(T_w - T_c) \quad (3)$$

$$h_1 = h_{rw} + h_{cw} + h_{ew}$$

Energy balance of condensing cover

$$h_1(T_w - T_c) = h_c(T_c - T_a) \quad (4)$$

The heat transfer coefficients will be given by

$$h_{cw} = 0.884\left[(T_w - T_c) + \frac{(P_w - P_c)(T_w + 273)}{268.9 \times 10^{-3} - P_w}\right]^{1/3} \quad (5)$$

$$h_{ew} = 16.273 \times 10^{-3} h_{cw} \frac{(P_w - P_c)}{(T_w - T_c)} \quad (6)$$

$$h_{rw} = \varepsilon_{eff}\sigma[(T_w + 273)^2 + (T_g + 273)^2] \quad (7)$$

$$P_w = \exp[25.317 - \frac{5144}{T_w + 273.15}] \quad (8)$$

$$P_c = \exp[25.317 - \frac{5144}{T_c + 273.15}] \quad (9)$$

$$\varepsilon_{eff} = (\frac{1}{\varepsilon_c} + \frac{1}{\varepsilon_w} - 1)^{-1} \quad (10)$$

From Eqs. 1 to 4, we have

$$\frac{dT_w}{dt} + aT_w = f(t) \quad (11)$$

$$a = \frac{U_L}{M_w c_w} \text{ and } f(t) = \frac{(\alpha\tau)_{eff} I + U_L T_a}{M_w c_w}$$

$$U_L = U_{wc} + U_{wg2} \quad (12)$$

$$U_{wc} = [\frac{1}{h_w} + \frac{1}{h_c}]^{-1} \quad (13)$$

$$U_{pg2} = [\frac{1}{h_{rpg2}} + \frac{1}{U_r}]^{-1}$$

$$U_{wg2} = [\frac{1}{h_{cpw}} + \frac{1}{U_{pg2}}]^{-1}$$

$$(\alpha\tau)_{eff} = \tau_{g1}\tau_{g2}r^{N}\alpha_p\frac{U_{wg2}}{U_{pg2}}$$

Solution to Eq. 5 will be,

$$T_w = \frac{f(t)}{a}[1 - \exp(-at)] + T_{w0}\exp(-at)$$

T_{w0} = inlet water temperature at t = 0

The temperature of condensing cover will be given by,

$$T_c = \frac{h_w T_w + h_c T_a}{h_w + h_c} \tag{14}$$

The hourly distillate output from solar still will be given by

$$m_{ew} = \frac{h_{ew}(T_w - T_c)}{L}3600kg/m^2h \tag{15}$$

The rate of heat rejected to room will be given by,

$$Q_r = h_c A(T_c - T_R)$$

In this work the room temperature is assumed equal to ambient temperature $T_R = T_a$ and

$$h_c = 5.7 + 3.8v$$

2.1 Room Temperature Gain

As the heat is rejected into the room, the temperature of room will be given by,

$$T_{r2} = \frac{Q}{M_a c_a} + T_{r1} \tag{16}$$

Q is the total heat rejected by condensing cover in room in Joules, $M_a c_a$ is the total heat capacity of room and T_{r1} is the initial room temperature.

3 Numerical Computations

Numerical computations have been performed to determine the performance of the proposed system for the climate of Delhi, India (28.7041°N, 77.1025°E). The climate of Delhi is characterized by cold winters and hot summers. There is always requirement of room heaters during winters. Generally majority of population uses either electric air heaters or gas-based heating systems to fulfil heating load requirements during winters. In present work author proposed a system that gives hot air in winters in addition the system provides pure drinking water throughout the year. The effect of varying the depth of water in basin of solar still has been investigated by performing numerical analysis for a day in winter season (Jan.). The hourly average of ambient temperature and solar radiation is given in Fig. 2. Other relevant parameters used for computations is shown in Table 1.

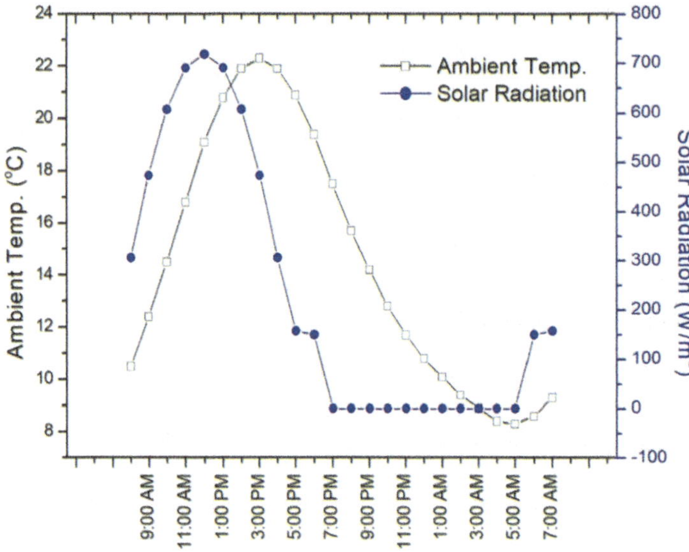

Fig. 2. Hourly average ambient temp. And solar radiation for Delhi (Jan.)

Table 1. Relevant parameters used for computations

$\tau_{g1} = 0.9$	$\tau_{g2} = 1.0$	$\alpha_p = 0.9$	$A = 1m^2$
$h_{cpw} = 100W/m^2C$	$U_r = 5.15W/m^{2o}C$	$r = 0.9$	$N = 2.5$
$c_w = 4190J/kg^oC$	$M_w =$ $10kg, 30kg, 50kg, 70kg$	$L = 2022 \times 10^3 J/kg$	$v = 2m/s$
$M_a = 28.3m^3$ $(10' \times 10' \times 10' room)$	$\rho_a = 1.204kg/m^3$	$c_a = 1005J/kg.K$	

3.1 Effect of Depth of Water

To investigate the effect of depth of water on the distillate output produced by still and the rate of heat rejected to room. Numerical computations have been performed by varying the depth of water in basin of solar still. The depths considered are 1cm, 3cm, 5cm, 6cm, 7cm and 10cm, the other relevant parameters used for computations are given in Table 1. To determine the hourly yield by still the initial water and temperature of condensing cover is taken equal to ambient. Different temperature dependent parameters have been computed by initial assumed values. The hourly yield is computed for twenty-four hours in a cycle, till the cycle repeats itself. It is seen from Fig. 3 that with increase in water depth the daily distillate output produced by still as well as the total daily heat rejected by still reduces. It is seen that 1cm depth is giving more daily yield but in off sunshine hours the yield almost becomes negligible however the total yield is maximum when the depth of water in basin is 1cm. By increasing the depth of water from 1cm to 10cm the decrease in daily distillate output is about 44.1% and the decrease in the total daily heat rejected is about 29.0%. However, in winters we are more concern about the heat rejected by the system during off sunshine hours as the room heating is more needed during off sunshine hours.

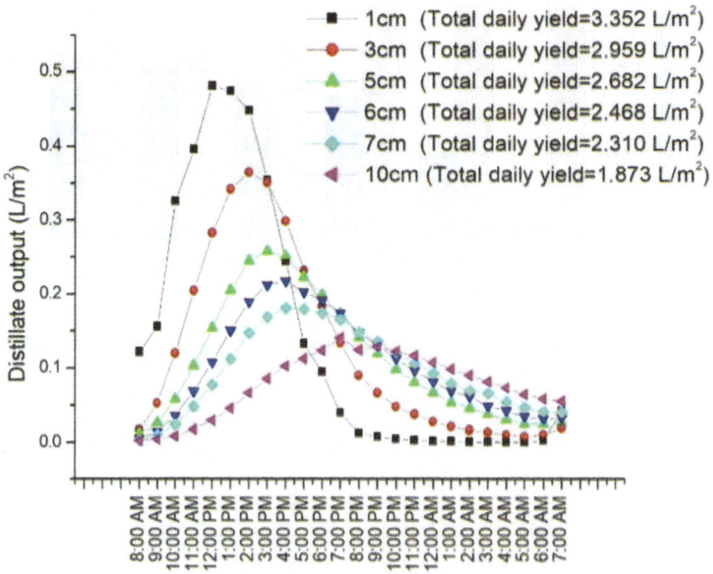

Fig. 3. Effect of water depth on the distillate output produced by still

To analyse this, we plotted the daily heat rejected from system during day and night for winter (Jan.). The result is as expected although there is a decrease in the amount of distilled output produced and the amount of heat rejected by increasing water depth but the heat released during night time is more at higher water depths due to the high heat capacity of basin water at higher depths (Fig. 4).

It is seen from Fig. 4 that the rate of heat energy released into the room is more during off sunshine hours for the depth of 10 cm but the yield produced will be less for this depth. To select the best possible depth of water in basin we analyzed by computing the final room temperature with different depths for a day in January month (Eq. 16).

It is seen from Fig. 5, at higher water depth of 10 cm the room temperature is above the desired room temperature range (about 20 °C) for 18 to 19 h in day for the depths 7cm–6cm the room temperature is in comfortable range for about 15–16 h in a day. Technically depth of 10 cm is giving best possible arrangement needed to heat room throughout the day but the distillate output produced will be less. For further computations the depth of 6cm during winters (Oct., Nov., Dec, Jan., Feb.) has been considered that gives the room comfort almost throughout the day with reasonable amount of water production. However, when room heating is not required to get higher yield the depth of water in basin is considered to be 1cm for summers and other months except winter months.

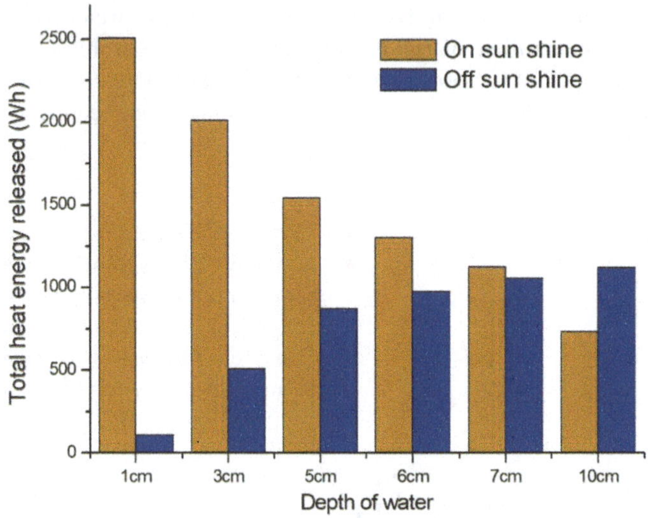

Fig. 4. Heat rejected during sunshine and off sunshine hours at different depths

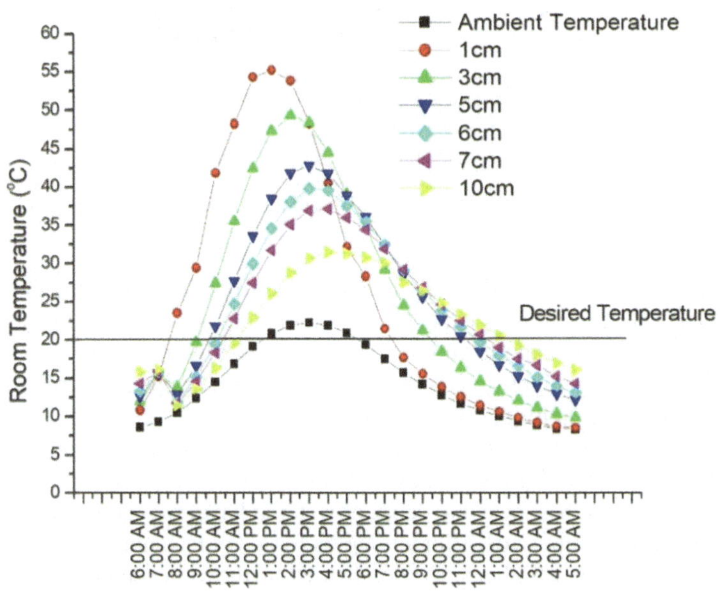

Fig. 5. Final room temperature after space heating by heat rejection

3.2 Annual Distillate Output Produced and the Heat Released for Space Heating by the Proposed System

The daily/monthly distillate output and the total heat rejected from the proposed system has been computed for the climate of Delhi. The total number of clear days taken as

291 days. It is seen that the maximum monthly output produced is about 165.8 L/m^2 for the month of May and minimum about 51.9 L/m^2 for the month of January (Fig. 6).

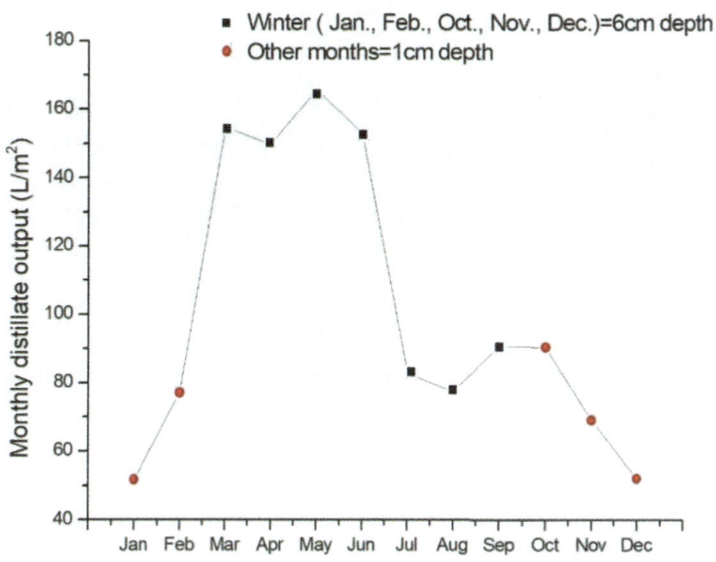

Fig. 6. Monthly distillate output produced by the proposed system for Delhi

The total monthly heat rejected has been computed only for winter months where the room heating is required. The heating months taken as Oct., Nov., Dec., Jan. and Feb. The monthly heat rejected into the room has been computed at 5cm water depth and shown in Fig. 7.

It is seen that the monthly heat rejected is maximum for the month of Oct. 75.9 kWh and minimum about 47.5 kWh in Jan. Total annual heat released for the space heating and distillate output produced by the proposed system will be 305.4 kWh and 1215 L/m^2 (Fig. 8).

3.3 Evaluation of Amount of Coal that Has Been Saved from Burning for Space Heating Using Proposed System

It has been seen the total amount of heat annually released for space heating that is produced from the proposed system is 305.4 kWh. Thus, not only compatible amount of coal but also related amount of carbon dioxide has also been saved from burning and releasing into atmosphere.

kW = 1 kJ/s = 1000 J/s; 1 kWh = 3600 X 1000 J/s = 3600 kJ/s

The calorific value of Indian coal very between 3345 kcal to 4703 kcal, taking average 4024 kcal gives value as 17000 kJ/kg

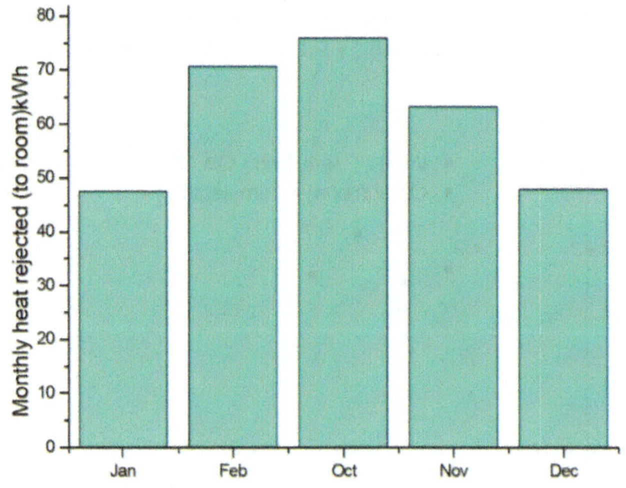

Fig. 7. Monthly heat rejected (kWh) by the proposed system for space heating in winter season for Delhi

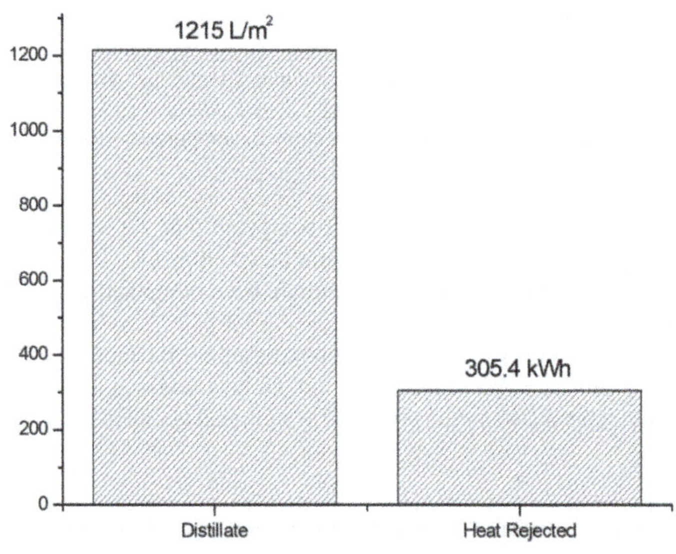

Fig. 8. Annual distillate output and heat rejected to room from proposed system

Normal efficiency of thermal power plant is 32.8%, thus for producing 3600 kJ/s of electricity (3600/0.328 = 10976 kJ/s should be liberated by burning the coal. Thus, amount of coal for 1 kWh electricity is required (10976/17000) = 0.6456 kg,

The amount of coal saved from burning by proposed system is 303.4 × 0.6456 = 197.18 kg ~ 200 kg

Indian Bituminous coal, on the other hand, has an average carbon content of approximately 66%. Since 66% of 1 kg is 660 g, there are about 660 g of carbon atoms in 1 kilo of bituminous coal and 12 g of carbon after complete combustion generates 44 g of carbon dioxide. Hence burning of 1 kg coal will generate = (44/12) × 660 = 2420 g = 2.420 kg

The amount of carbon dioxide prevented to be liberated in earth atmosphere will be = 200 × 2.420 = 484 kg of CO_2 will not be released in earth atmosphere.

4 Conclusions

In this work authors proposed an inverted absorber solar still which can be utilized for generating pure water additionally it can be utilized for space heating. Water evaporates from basin due to reflected solar radiation received from the bottom condenses into the inner side of copper condensing cover as the vapor condenses the latent heat of condensation is given to the cover which is further rejected unto the space to be heated.

In this work authors suggested two different strategies for summer and winters, during winter season when room heating is required the heat from the condensing cover is released into the room for heating, whereas during summer season when the room heating is not required the heat from condenser plate is taken away from the room to outside atmosphere through a maneuverable portable ducting arrangement.

In cold and composite climate receiving reasonable amount of solar radiation there is always need to heat the room during winters which is generally taken either by electric heaters, in present work we proposed a system in which inverted absorber type of solar still is installed in the floor area of the room. The heat rejected by solar still into the room thus helps to heat the room thus reduces the amount of electric energy required for room heating at the same time generating distill water. It is seen that for 1m2 system in the climate of Delhi the annual distillate output produced by the proposed system is about 1215 L/m2 and the total amount of heat rejected into the room will be 305.4 kWh. This way about 200 kg of coal and 484 kg of CO2 has been prevented from burning and liberated in to atmosphere respectively for a space size of 3.05 m × 3.05 m × 3.05 m.

In present work numerical analysis has been performed by considering $1m^2$ size of the system however one can increase the total size of the inverted still installed in the room depending on the size available and the total heating load requirements.

Nomenclature

h_{rpg2}	Radiative heat transfer coefficient from glass cover to plate	W/m^2 °C
T_p	Temperature of plate	°C
T_{g2}	Temperature of glass cover 2	°C
T_a	Temperature of ambient	°C
U_r	Overall heat transfer coefficient from bottom plate	W/m^2 °C
I	Solar Radiation	W/m^2

(continued)

(*continued*)

T_w	Temperature of water in basin	°C
T_c	Temperature of condensing cover	°C
h_{ew}	Evaporative heat transfer coefficient from water to condensing cover	W/m^2 °C
h_{cw}	Convective heat transfer coefficient from water to condensing cover	W/m^2 °C
h_{rw}	Radiative heat transfer coefficient from water to condensing cover	W/m^2 °C
P	Saturated vapour pressure	
α_p	Absorptivity of bottom plate	
τ_{g1}	Transmissivity of glass cover 1	
τ_{g2}	Transmissivity of glass cover 2	
r	Reflectivity of mirror	
N	Average number of reflections	

References

1. Somwanshi, A., Tiwari, A.K.: Performance enhancement of a single basin solar still with flow of water from an air cooler on the cover. Desalination **352**, 92–102 (2014)
2. Tiwari, A.K., Somwanshi, A.: Techno-economic analysis of mini solar distillation plants integrated with reservoir of garden fountain for hot and dry climate of Jodhpur (India). Sol. Energy **160**, 216–224 (2018)
3. Dhiman, N.K.: Transient analysis of a spherical solar still. Desalination **69**(1), 47–55 (1988)
4. Arunkumar, T., et al.: An experimental study on a hemispherical solar still. Desalination **286**, 342–348 (2012)
5. Sathyamurthy, R., Kennady, H.J., Nagarajan, P.K., Ahsan, A.: Factors affecting the performance of triangular pyramid solar still. Desalination **344**, 383–390 (2014)
6. Suneesh, P.U., Jayaprakash, R., Arunkumar, T., Denkenberger, D.: Effect of air flow on "V" type solar still with cotton gauze cooling. Desalination **337**, 1–5 (2014)
7. Gad, H.E., El-Din, S.S., Hussien, A.A., Ramzy, K.: Thermal analysis of a conical solar still performance: an experimental study. Sol. Energy **122**, 900–909 (2015)
8. Tiwari, G.N.: Solar energy: fundamentals. Modeling and Applications, Narosa Publishing House, New Delhi, Designs (2002)
9. Abdullah, A.S.: Improving the performance of stepped solar still. Desalination **319**, 60–65 (2013)
10. Tiwari, G.N., Suneja, S.: Performance evaluation of an inverted absorber solar still. Energy Convers. Manage. **39**(3–4), 173–180 (1998)
11. Suneja, S., Tiwari, G.N.: Effect of water depth on the performance of an inverted absorber double basin solar still. Energy Convers. Manage. **40**(17), 1885–1897 (1999)
12. Suneja, S., Tiwari, G.N.: Parametric study of an inverted absorber triple effect solar still. Energy Convers. Manage. **40**(17), 1871–1884 (1999)

Silicon Dots in Matrix of Aluminum-Doped Zinc Oxide as a Wide-Energy Bandgap Absorber for Solar Cells

Thipwan Fangsuwannarak[1](✉), Supanut Laohawiroj[1], Peerawoot Rattanawichai[1], and Karoon Fangsuwannarak[2]

[1] School of Electrical Engineering, Institute of Engineering, Suranaree University of Technology, Nakhon Ratchasima, Thailand
thipwan@g.sut.ac.th

[2] School of Mechanical Engineering, Institute of Engineering, Suranaree University of Technology, Nakhon Ratchasima, Thailand

Abstract. A low-cost spin coating technique and rapid thermal annealing (RTA) methods have been utilized for the preparation of aluminum-doped zinc oxide (AZO) with a Si dots thin film on the quartz substrates. The dispersed Si dots particle in the film showed that these additional Si dots can be tunable material for light absorption applied in photovoltaic work. X-ray diffraction was used to study the nanocrystalline silicon and zinc oxide with the crystal grain size of ~ 100 and ~ 20 nm presented in the films, respectively. The optical result of the thin films shows that $< 20\%$ reflectance and $> 80\%$ transmittance were presented in the film. The absorption edge of thin films that have occupied Si dots was expanded from the original zinc oxide which strongly absorbs UV wavelengths (~380 nm) to near-infrared wavelength (~800 nm) depending on the amount of Si dots layer. For the electrical properties of the film, the two-probe method was used to observe the photocurrent gain under the illumination compared with the dark test and showed that the film can greatly respond to the light and provide higher current density values. Additionally, the increase in annealing temperature as RTA and the amount of zinc oxide layer leads to the enchantment of the current density. Hence, these zinc oxide thin films can be an alternative candidate for the tunable absorption emitter layer of solar cells and can be applied to optoelectronic devices.

Keywords: zinc oxide doped aluminum · silicon powder · thin film

J. Baeyens et al. (Eds.): ESRE 2023, pp. 137–151, 2025.
https://doi.org/10.1007/978-981-96-4345-5_11

1 Introduction

The silicon heterojunction (SiHJ) solar cell is one type of solar cell that has amorphous silicon (a-Si) consisting in its structure between the metal contact and crystalline silicon. It is used as the passivation layer that can reduce carrier recombination and increase the fill factor [1]. The passivation layer on SiHJ fabrications can be carried out by plasma-enhanced chemical vapor deposition (PECVD) [2–4] with high efficiency of up to 24.51% [5]. The a-Si for SiHJ can be acted as the emitter layer, due to the doping concentration of phosphorus or boron atom [6]. The emitter is the area of Si solar cells that can be mentioned as the heavily doped region. This region can inject or emit most of the carriers both electrons and holes because it can reduce the carrier recombination, reduce the contact resistance [7], collect the photogenerated carriers, and improve the internal quantum efficiency for blue wavelength [8].

Solar cells are semiconductor devices that can directly convert sunlight to electric power, most of them are fabricated from silicon (Si) material which has an energy band gap of 1.1 eV and the Si cell can provide maximum efficiency of 30% due to the detailed balance limit of efficiency [9]. The concept of tandem cells can be applied to be on top of the solar cell to improve light absorption and efficiency because the different band gaps that are higher than Si can convert a narrow range of light energies to their energy gap [10].

Zinc oxide (ZnO) is well known for applying much attention to research due to the direct band gap of n-type material and its optical properties, it has a broad energy band of 3.37 eV, which can especially respond to the ultraviolet (UV) wavelength [11]. The transparency of ZnO film can provide a higher wavelength of light passing through the film easily. The ZnO thin film process has been carried out by many methods such as sputtering [12], plasma-enhanced chemical vapor deposition [13], spray pyrolysis [14], and the sol-gel method [15]. In addition, the ZnO can be a composite material with other materials to improve their electrical properties in supercapacitor applications [16], to enchant the thermal conductivity and dielectric constant [17], and to be the ferroelectric ZnO nanowire polymer composite layer for instantaneous dynamic motion sensor [18]. Moreover, our previous work with nanocrystalline ZnO doped with Bi films and Si powder in the matrix of ZnO doped Bi has also been reported about their properties and emitter layer on solar cells [19–22].

In this work, we have improved and studied the AZO thin film and AZO with Si powder thin film called Si dots from waste silicon wafers. The comparison between normal thermal annealing (TA) of 550 °C for 2 h and rapid thermal annealing of 930 °C for 30 s has been reported. The Si dots thin films convinced the improvement in the electrical properties and absorbance mechanism.

2 Experimental Details

2.1 Material

Si powder as a material for Si dot thin films was obtained from waste Si wafers transmogrified by grinding technique. The average small particles of Si powder of 4 μm were accomplished by conventional ball mill planetary QM2L which was detailed in our previous work [19, 23]. The substances for synthesizing the AZO consist of zinc acetate dehydrate (ZnAc, Zn-$(CH_3COO)_2 \cdot 2H_2O$), 99.5%) as a ZnO source, monoethanolamine (MEA, $NH_2C_2H_4OH$, 99%), aluminum chloride hexahydrate ($AlCl_3$, $AlCl_3 \cdot 6H_2O$) as an aluminum dopant source, ethanol (EtOH, CH_3CH_2OH), isopropanol (IPA, $(CH_3)_2CHOH$).

2.2 ZnO Thin Film and Si Dots Thin Film Preparation

Aluminum-doped zinc oxide (AZO) was prepared solution as described in this section. Firstly, ZnAc of 0.2 M was dissolved in a solution of IPA-MEA and kept the ratio of MEA:Zn equal to 0.75:1, then, the solution was stirred at 60 °C for 30 min. After that, the 0.2M of $AlCl_3$ dissolved in EtOH was added to the ZnO-IPA-MEA solution to receive 1% atom of Al. Then, the solution continued stirring at 60 °C for 30 min and the clear solution was aged at room temperature for 24 h. For silicon dots solution preparation, AZO solution was utilized as a matrix for Si powder. The ratio of the Si powder to AZO solution was kept at 1 g:66.67 ml and well dispersed in an ultrasonic bath for 10 min. After that, the spin coating technique was used for depositing the thin film on quartz substrates with a 2-step coating at 100 rpm for 15 s and 2500 rpm for 30 s, respectively. Thin films were dried at 200 °C for 30 min and then some thin films were repeated deposition and dried to obtain 3 layers of AZO thin film. The difference annealing of the film was separated to be thermal annealing (TA) at 550 °C for 1 h. and rapid thermal annealing (RTA) at 930 °C for 30 s.

2.3 Characterization

The prepared AZO and Si dots thin film were measured the thickness by an optical profiler (Bruker) and characterized the surface morphology, crystalline structure, and crystal grain size of ZnO and Si, optical properties, and electrical properties by field emission scanning electron microscopy ((FESEM), Auriga Carl Zeiss), X-ray diffractometer (Bruker D8 corporation), UV–VIS–NIR spectrophotometer (LAMBDA 950 of PerkinElmer Inc.,) and 2400 electrometer (Keithley), respectively.

3 Result

3.1 Surface Morphology

Figure 1 shows the 20k magnification of surface morphology of the AZO and Si dots thin films in different annealing temperatures of 550 °C and 930 °C and in different layers of 1 and 3 layers on the quartz substrate. In Fig. 1(a) and (b), FESEM images show the surface morphology of the AZO thin film of 1 layer with TA, (AZO_1L_TA) and RTA (AZO_1L_RTA), they showed that the ZnO particles cover the entire area on the quartz substrate. Figure 1(c) of the AZO thin film of 3 layers with RA (AZO_3L_RTA) expressed that the ZnO particles were homogeneously agglomerated and inset showed the nano spherical grain of AZO with the size of 8–23 nm. For Si dots thin films of AZO thin film of 1 layer with TA, (AZO_Si_1L_TA) and RTA (AZO_Si_1L_RTA) and 3 layers with RTA (AZO_Si_3L_RTA) were shown in Fig. 1(d), (e) and (f), respectively. The Si dots thin films appear the Si particles covered by nano spherical ZnO and scattered on the film with Si size less than 2 μm, however, the Si dots thin films with 3 layers show the agglomeration of Si. The average thickness of the thin films in Table 1 exhibits that the AZO thin film for 1 and 3 layers have an average thickness of 65 and 120 nm, respectively, while the Si dots films of 1 and 3 layers have an average thickness of 120 and 170 nm, respectively.

Figure 2 shows the XRD pattern of prepared AZO and Si dots thin films. XRD pattern of Si corresponded to a reference pattern of Si crystalline structure (JCPDS 00-027-1402) which is composed of 2θ around 28.4°, 47.3°, and 56.1° with orientations along (111), (220), and (311) planes, respectively. It is noted that the intensity of the crystalline Si peak increases due to the amount of possessed Si in the film. ZnO peaks were observed at positions about 31.8°, 34.5°, and 36.5° with orientations along (100), (002), and (101) planes, respectively, which correspond to a reference pattern of spherical ZnO crystal (JCPDS 36–1451). ZnO peak at 2θ between 30°–40°, the intensity of Si dots thin films with 1 layer expresses the 3 tiny peaks of ZnO, while the 3 layers of Si dots thin films showed the higher peaks, this may indicate that the AZO hardly detect the peak for small thickness, in addition, the increase in peak intensity can explain the improvement in the crystallinity of the film [24]. Moreover, the XRD pattern of ZnO of Si dots films did not appear in the amorphous phase as Al_2O_3 and $ZnAl_2O_4$ because of the high-temperature annealing and Al dopant, which indicates that the Al atom does not change the crystal structure of ZnO [25, 26].

3.2 Structural Property

The simple cubic structure of Si and hexagonal wurtzite structure of ZnO of (111) and (002) crystallographic planes, respectively, can be verified as the average crystal grain size by Scherrer's relation equation in Eq. (1), and the calculated results are presented in Table 1. It showed that the average crystal grain size of ZnO of AZO with RTA is bigger than TA, it is noted that the high-temperature annealing can increase the size of ZnO crystal [27]. For the Si dots film, it is shown that the crystal grain size of ZnO decreases both RTA for 1 and 3 layers. The average crystal grain size of Si dot thin films of AZO_Si_1L_TA thin film is smaller than AZO_Si_1L_RTA and AZO_Si_3L_RTA,

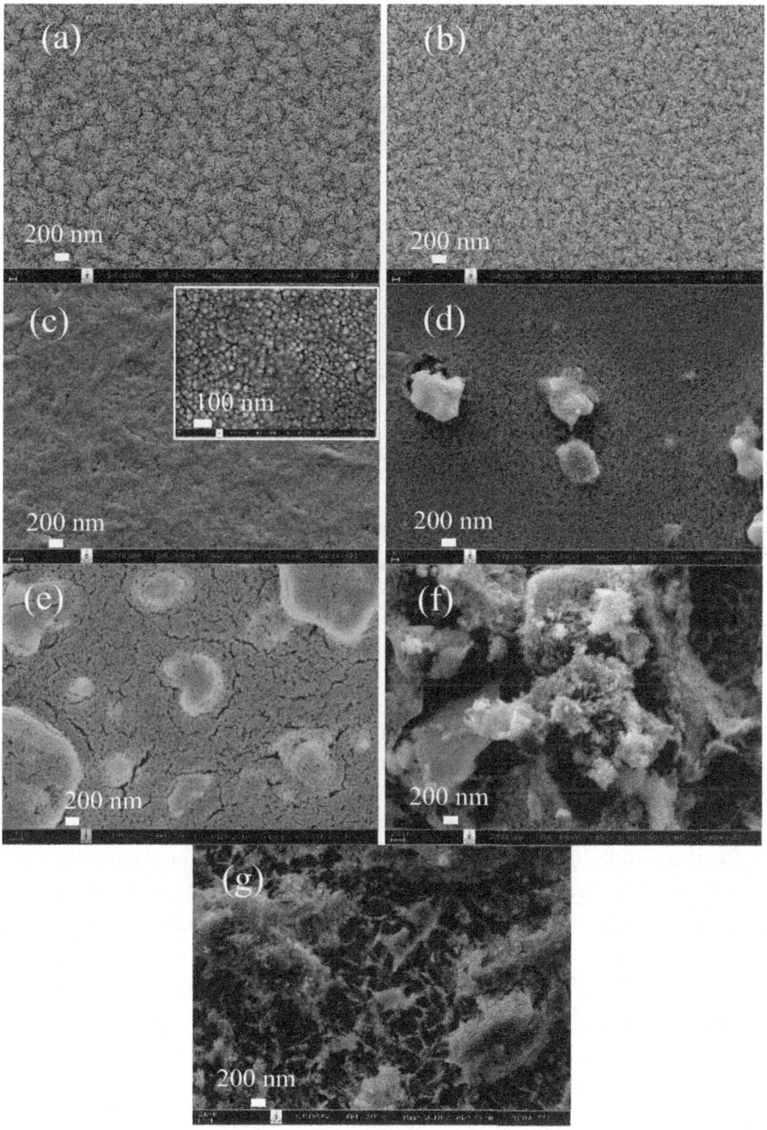

Fig. 1. FESEM images of AZO and Si dots thin films of (a) AZO_1L_TA, (b) AZO_1L_RTA, (c) AZO_3L_RTA, (d) AZO_Si_1L_TA, (e) AZO_Si_1L_RTA, (f) AZO_Si_3L_RTA and (g) AZO_SI_AZO RTA

this probably implied that Si powder can reduce the crystal grain size of the matrix of ZnO.

$$d = \frac{0.9\lambda}{Bcos(\theta_B)} \tag{1}$$

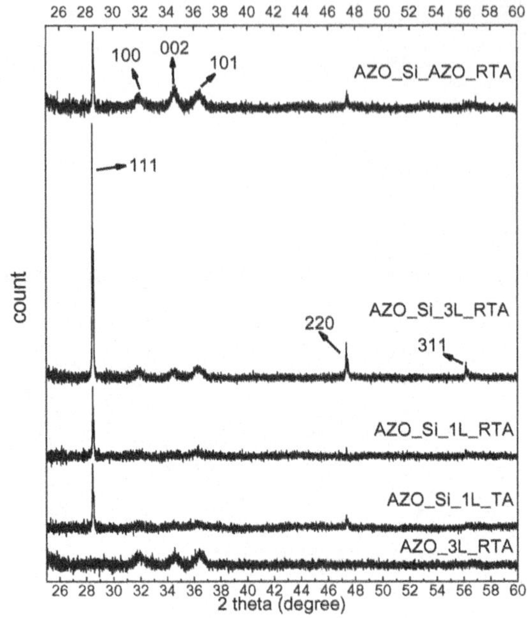

Fig. 2. XRD pattern of Si dots thin films.

where d is the average crystal grain size (nm), λ is the X-ray radiation wavelength with 0.154 nm, B is the full-width half maximum (radian), and θ_B is the Bragg diffraction angle of XRD peak (degree)

The small shift of (111) peak and (002) broad peak refers to the change of lattice constants for the simple cubic structure of Si and the hexagonal wurtzite structure of ZnO, respectively. The strain along the c-axis (ε) was determined to evaluate the change of lattice constants given in Eq. (2) and the results are reported in Table 1. The result of strain along the c-axis expresses the compressive strain (negative sign) that occurs for every sample both AZO thin film and Si dots thin films, it is due to the interstitial impurity doping atom and annealing temperature [21].

$$\varepsilon = \frac{C - C_0}{C_0} \tag{2}$$

where C is the lattice parameter of the film calculated from XRD data, C_0 is the lattice parameter of the pure Si crystal (C_0 = 5.43 Å) [28] and ZnO crystal (C_0 = 5.207 Å) [29]

3.3 Electrical Properties

Figure 3(a) shows the J-V curve from the 2-probe method under the dark and illumination conditions. The linear curve expresses the ohmic behavior between the heterojunction of the metal and thin film. The different carrier generation under the illumination test was exhibited in Fig. 3(b), and the current density of AZO thin films increases. When

Table 1. Crystalline size and strain along the c-axis of Si and ZnO

Condition	Thickness (nm)	Crystal grain size (nm)		Strain along c-axis	
		Si (111)	ZnO (002)	Si (111)	ZnO (002)
AZO_1L_TA	68.43 ± 15.81	–	11.09	–	−0.327
AZO_1L_RTA	60.57 ± 0.96	–	19.36	–	−0.014
AZO_3L_RTA	120.46 ± 8.13	–	20.36	–	−0.372
AZO_Si_1L_TA	124.36 ± 9.65	90	11.84	−0.024	−0.054
AZO_Si_1L_RTA	130.29 ± 20.03	106	10.38	−0.103	−0.092
AZO_Si_3L_RTA	176.13 ± 10.59	107	15.89	−0.096	−0.392
AZO_Si_AZO_RTA	189.63 ± 18.72	73	23.20	−0.339	−0.567

they were exposed to the light, this Si powder in the Si dots thin film did not lead to light-induced degradation due to the increment of trap state between the Si powder and AZO matrix [30]. The film can generate the carrier and increase the conductivity under the illumination condition as shown in Fig. 3(c) with the result of the photocurrent gain value from the ratio of current density between illumination and dark test. The relationship of the conductivity (σ) was calculated in Eq. (3). The 3 layers of AZO thin film provide high values of conductivity if compare with 1 layer of AZO because the electron mobility was improved from the increase of ZnO crystal grain size. The reduction of the grain boundaries in the film and the improvement of the film surface morphology were found [31]. The Si dots films express smaller conductivity than AZO thin film because the interface between Si and ZnO occurs in the small trap state.

$$\sigma = \frac{J \times l}{V} \tag{3}$$

where V is the applied voltage (V), J is the current density (A/cm^2) and l is the distance between the metal contact that applied voltage (cm)

3.4 Optical Properties

Figure 4(a) and (b) express the light reflection and light transmission spectra in wavelength of 200–1500 nm for the different conditions of AZO and Si dots thin film, respectively. It is shown that the reflectance values of all samples are less than 15% in all regions, except the AZO_Si_3L_RTA which has a value of reflectance of more than 15% in the infrared region. The transmittance of the AZO and Si dots thin film of 1 layer expresses the light can transmit through more than 80% in the visible and infrared regions. Note that the increase of Si powder in the film can conceal the film and decrease the transmittance in all wavelengths.

The optical absorption coefficient can be calculated by the relationship from Eq. (4) and can be plotted in the curve as shown in Fig. 5. Note that the Silicon dots thin film is a composite material of Si powder in ZnO matrix, hence the optical bandgap can be

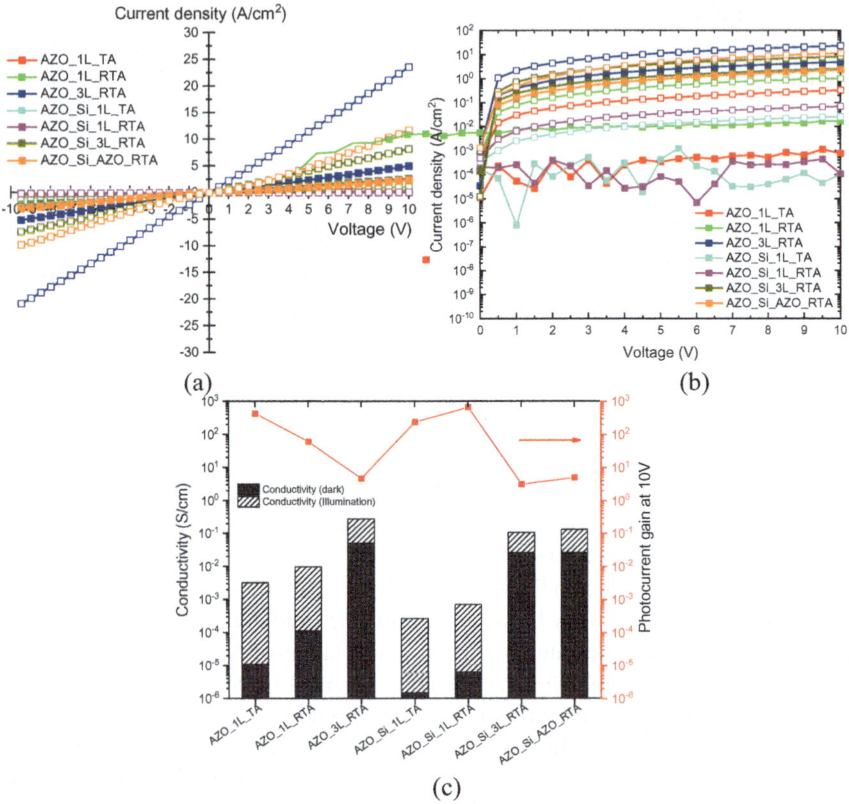

Fig. 3. JV curve of thin films in (a) linear and (b) semi-log scale. And (c) the curve of resistivity and photocurrent gain at 10 V of thin films.

implied as the estimation of the absorption edge [32].

$$\alpha = \frac{1}{t} ln\left(\frac{\sqrt{(1-R)^4 + 4T^2R^2} - (1-R)^2}{2RT}\right) \tag{4}$$

where t is the average thickness of the film (cm)

R is the reflectance value.

T is the transmittance values.

The 1 layer of AZO thin film with TA and RTA has an absorption edge of 3.3 eV, whereas the 3 layers of AZO thin film can be slightly shifted to 3.2 eV. This indicates that a group of these AZO thin films efficiently absorb UV light. In addition, the Si dots films observed that the absorption edge can be expanded to the visible region and near-infrared region depending on the number of Si dots layers.

Moreover, the refractive index curve is shown in Fig. 6(a) can be computed from Eq. 5 and the result at the wavelength 600 nm is reported in Table 2.

$$R = \frac{(n-1)^2 + k^2}{(n+1)^2 + k^2} \tag{5}$$

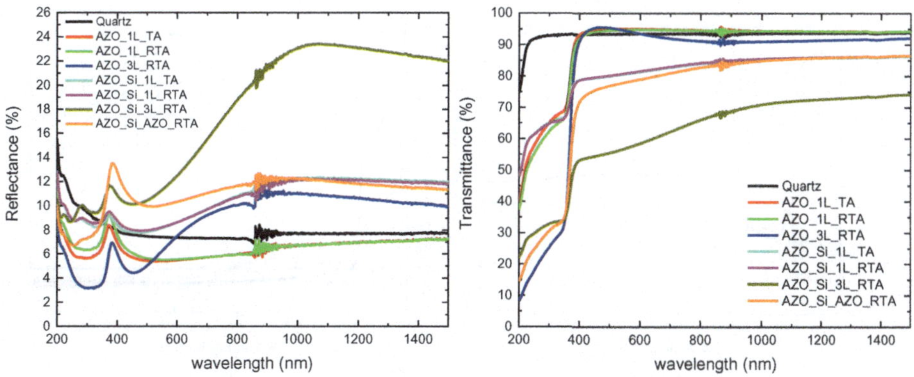

Fig. 4. (a) Reflectance and (b) transmittance spectra of AZO and Si dots thin films

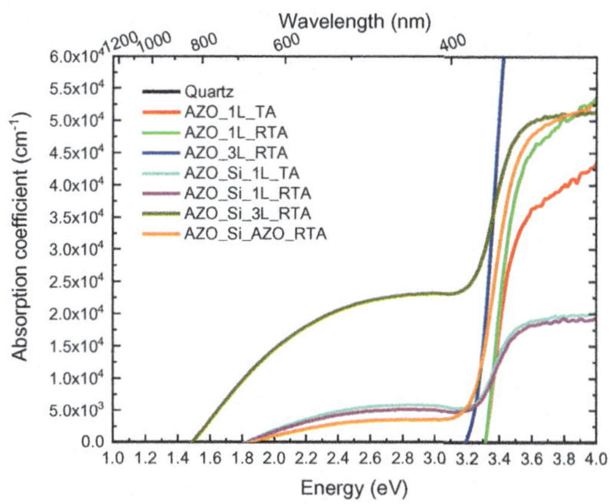

Fig. 5. Absorption coefficient of AZO and Si dots thin films

where n is the refractive index of the film and k is the extinction coefficient that can define as $k = \alpha/4\pi$

The result suggests that an increase in the number of layers of the film results in an increase in refractive index and the refractive values are less than 3. Thus, this film can be applied to be as an anti-reflection coating (ARC) layer and emitter layer for solar cell application. Figure 6(b) shows the extinction coefficient of the AZO and Si dot film. It is observed that the AZO thin film with 1 layer has an extinction coefficient lower than 3 layers of AZO thin film and in the UV wavelength region. Nevertheless, the wavelength is more than 380 nm, it appears that the extinction coefficient becomes negative due to calculation. For the Si dots thin film, it is observed that the value of extinction value is dramatically decreasing in the UV region, moreover. The positive value of the extinction coefficient can be expanded to the visible region.

Dielectric constant can be separated into 2 parts the real dielectric or dielectric constant ($\varepsilon\prime$) and imaginary dielectric or dielectric loss (ε'') as a function of wavelength which

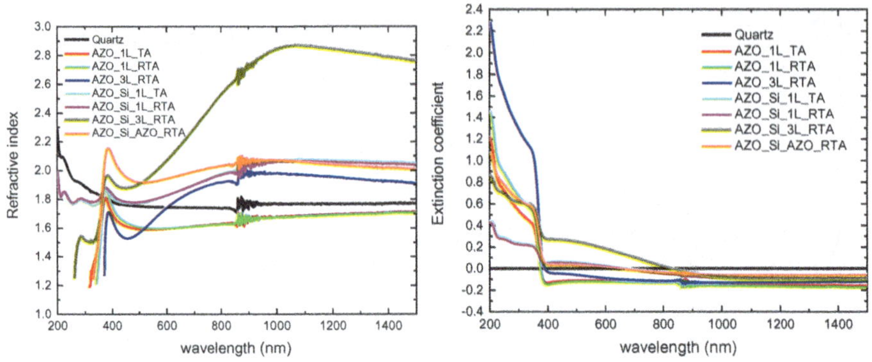

Fig. 6. (a) Refractive index and (b) extinction coefficient of AZO and Si dots thin films

are performed in Fig. 7(a) and (b) and can calculate from Eqs. (6) and (7), respectively.

$$\varepsilon\prime = n^2 - k^2 \tag{6}$$

$$\varepsilon'' = 2nk \tag{7}$$

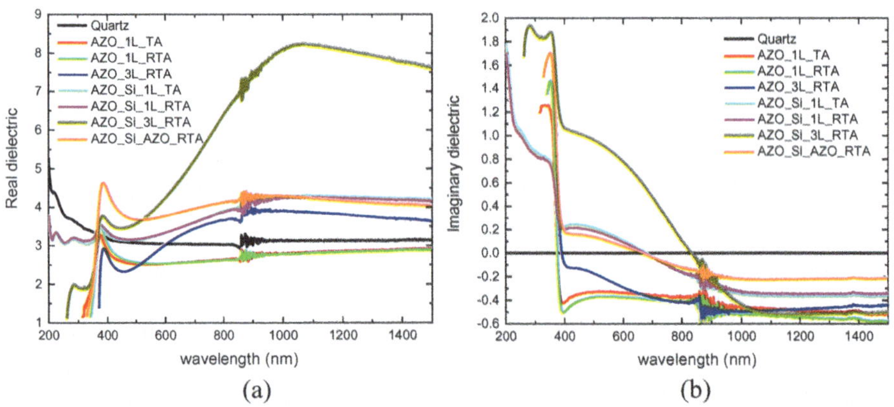

Fig. 7. (a) real dielectric and (b) imaginary dielectric of ZnO and Si dots thin films

Moreover, the loss tangent or dielectric loss factor (tan "δ"), which is the ratio of the imaginary dielectric to the real dielectric shown in Eq. (8)

$$tan\delta = \frac{\varepsilon''}{\varepsilon\prime} \tag{8}$$

The value of real dielectric expresses the dispersion of the electromagnetic waves traveling inside the materials. The imaginary dielectric is related to the energy absorption from the electric field from dipole motion. It can present the indication of how well a material dissipates stored energy into the heat [33, 34].

From the real dielectric report in Table 2, it is found that, at the wavelength of 600 nm, the real dielectric value increase with the thickness of the layer increasing both AZO and Si dots thin film because of the quantum confinement effect of the zinc oxide grain crystal [35]. This can also occur in the Si dot thin film as well. It is possibly due to the quantum confinement effect of the Si grain crystal. From Eqs. (7) and (8), the imaginary dielectric and loss tangent at the wavelength of 600 nm in Table 2. Express the negative sign of the AZO thin film, while the sample of Si dots thin film showed the positive.

Table 2. Absorbance edge, refractive index and dielectric parameter at wavelength 600 nm of ZnO thin film and Si dots film

Name	Absorbance edge		Refractive index	Dielectric		Loss tangent
	nm	eV		Real	Imaginary	
AZO_1L_TA	374	3.315	1.609	2.556	−0.327	−0.128
AZO_1L_RTA	374	3.315	1.615	2.561	−0.366	−0.143
AZO_3L_RTA	389	3.190	1.747	3.053	−0.286	−0.094
AZO_Si_1L_TA	685	1.810	1.833	3.358	0.115	0.034
AZO_Si_1L_RTA	678	1.830	1.831	3.360	0.097	0.029
AZO_Si_3L_RTA	832	1.490	2.060	4.251	0.782	0.183
AZO_Si_AZO_RTA	672	1.845	1.949	3.799	0.068	0.018

The wavelength of the light can be converted to the frequency (f) form by computing the wavelength as the $f = c/\lambda$, where c is the speed of light. Hence, the real part, imaginary part and loss tangent of the dielectric can also be presented in terms of frequency dependence as shown in Fig. 8(a), (b), and (c), respectively.

The properties in the range of high frequency of 10^{13}–10^{15} Hz express the electronic polarization which involves a displacement of electron cloud relative to nuclei in the atom forming the molecule of the dielectric under applied field [36, 37]. Figure 8(d), (e), and (f) show the real dielectric, imaginary dielectric and tangent loss value of the film considering the frequency of 7.5, 5.00 and 3.84 ($\times 10^{14}$ Hz) at the same wavelength of 400 nm (UV), 600 nm (visible) and 780 nm (infrared), respectively. The AZO thin film with 1 layer depicts the increase of real dielectric as the increase of frequency, while the AZO thin film with 3 layers and Si dots thin film is opposite. For the trend of imaginary dielectric and loss tangent of both AZO and Si dots thin films are similar which are increase with frequency increase.

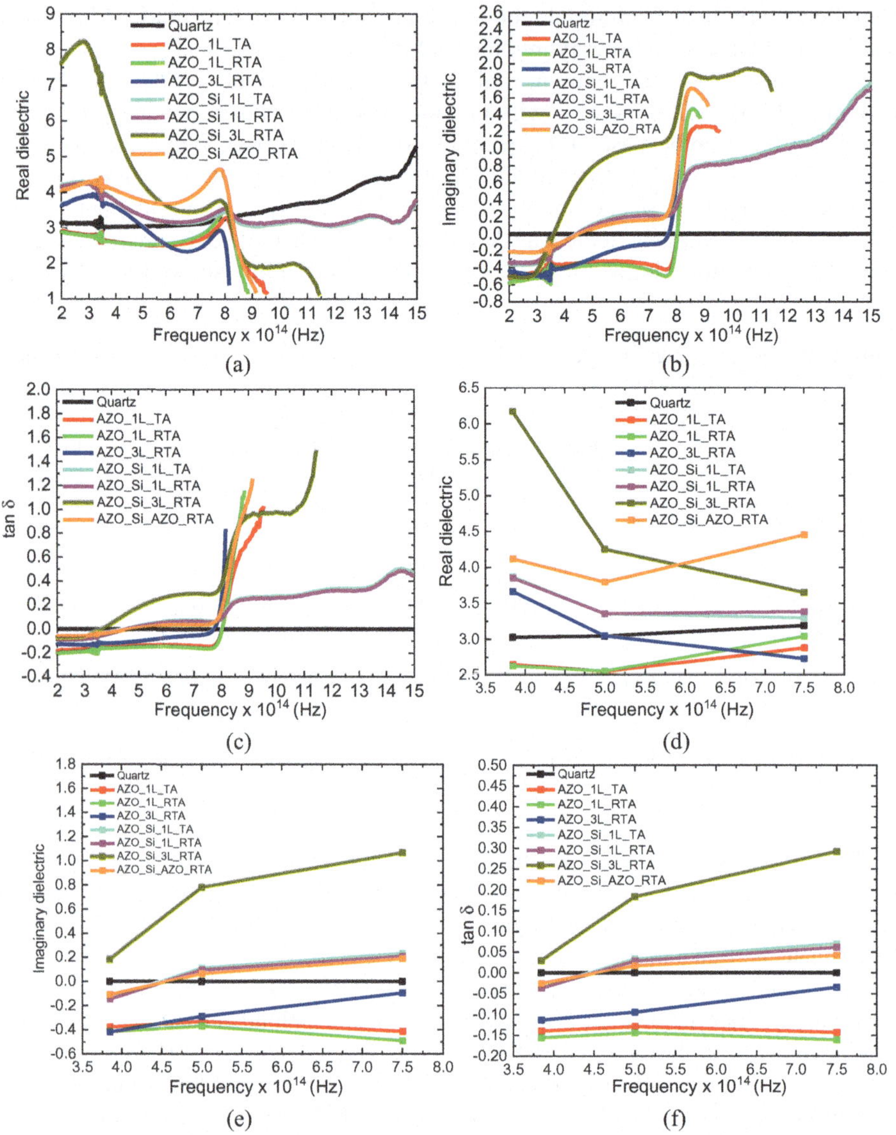

Fig. 8. (a) real dielectric (b) imaginary dielectric and (c) loss tangent of AZO and Si dots thin films in the function of frequency with the value of (d) real dielectric (e) imaginary dielectric and (f) loss tangent at frequency of 7.50, 5.00 and 3.84 ($\times 10^{14}$ Hz)

4 Conclusion

Si dots films have been successfully fabricated from Si ink which is synthesized from transmogrified waste Si wafer into Si powder with AZO solution. The simple method of spin coating technique has been used for thin film depositing on quartz substrate and the thin films were annealed with different processes of normal annealing and rapid thermal annealing. The surface morphology of the AZO thin film with RTA expressed the nano spherical grain size of ZnO. The improvement of the grain boundary of ZnO particles can be done with RTA and depositing the film more than 1 layer. The crystal structure of AZO thin films expresses the peak of ZnO from the XRD pattern with the crystal size between 10 – 20 nm depending on the temperature annealing, moreover, the XRD pattern of Si dots thin film nowhere appears in the amorphous phase. Although the films were annealed with a different method, they still attributed the compression strain. The AZO thin films have their properties to absorb the UV wavelength but the existence of Si powder in Si dots film can improve the energy absorption in visible wavelength that is reported from the absorption coefficient. The result from the illumination test with the two-probe method provides the photocurrent gain which means that the AZO and Si dots thin films are fewer defect states and disappear the light-induced degradation. Hence, the AZO thin films and Si dots films can be applied on the emitter layer that needs to improve the absorption properties of the solar cell.

Acknowledgements. This work has been supported by Suranaree University of Technology (SUT) Research and Development Fund, Thailand.

References

1. Christophe B., Stefaan D.W., Antoineet D. et al.: Chapter two—amorphous silicon/crystalline silicon heterojunction solar cells, in semiconductors and semimetals. Semicond. Semimet. **90**,73–120 (2014). Elsevier
2. Das, U.K., Burrows, M.Z., Lu, M., et al.: Surface passivation and heterojunction cells on Si (100) and (111) wafers using dc and rf plasma deposited Si: H thin films. Appl. Phys. Lett. **92**(6), 063504 (2008)
3. Kaining D., Aeberhard U., Finger F. et al.: Silicon heterojunction solar cell with amorphous silicon oxide buffer and microcrystalline silicon oxide contact layers. Phys. Status Solidi (RRL) —Rapid Res. Lett. **6**(5), 193–195 (2012)
4. Yoshikawa, K., Kawasaki, H., Yoshida, W., et al.: Silicon heterojunction solar cell with interdigitated back contacts for a photoconversion efficiency over 26%. Nat. Energy **2**(5), 17032 (2017)
5. Weiyuan, D., Andreas, L., Karsten, B., et al.: A route towards high-efficiency silicon heterojunction solar cells. Prog. Photovoltaics Res. Appl. **30**(4), 384–392 (2022)
6. Martín de Nicolás S., Muñoz D., Ozanne A.S. et al.: Optimisation of doped amorphous silicon layers applied to heterojunction solar cells. Energy Procedia **8**, 226–231 (2011)
7. Kyuwan, S., Bonggi, K., Hoongjoo, L., et al.: Selective emitter using a screen printed etch barrier in crystalline silicon solar cell. Nanoscale Res. Lett. **7**(1), 410 (2012)
8. Rahman, M.Z.: Status of selective emitters for p-type c-Si solar cells. Opt. Photonics J. **2**, 129–134 (2012)
9. Shockley, W., Queisser, H.J.: Detailed balance limit of efficiency of p-n junction solar cells. J. Appl. Phys. **32**(3), 510–519 (1961)
10. Green, M.A.: Third generation photovoltaics: solar cells for 2020 and beyond. Physica E **14**(1), 65–70 (2002)
11. Kołodziejczak-Radzimska, A., Jesionowski, T.: Zinc Oxide-from synthesis to application: a review. Materials (Basel, Switzerland) **7**(4), 2833–2881 (2014)

12. Ramadan, R., Dadgostar, S., Manso-Silvan, M., et al.: Silver-enriched ZnO: Ag thin films deposited by magnetron co-sputtering: post annealing effects on structural and physical properties. Mater. Sci. Eng. B **276**, 115558 (2022)
13. Barankin, M.D., Gonzalez, E., II., Ladwig, A.M., et al.: Plasma-enhanced chemical vapor deposition of zinc oxide at atmospheric pressure and low temperature. Sol. Energy Mater. Sol. Cells **91**(10), 924–930 (2007)
14. Grothe, R., Knust, S., Meinderink, D., et al.: Spray pyrolysis of thin adhesion-promoting ZnO films on ZnMgAl coated steel. Surf. Coat. Technol. **394**, 125869 (2020)
15. Rattanawichai, P., Fangsuwannarak, T., Phatthanakun, R., et al.: High photocurrent gain of spherical nano-crystalline ZnO: Bi film for advanced solar cells application. Chiang Mai J. Sci. **45**(5), 1995–2004 (2018)
16. Wang, Y., Xiao, X., Xue, H., et al.: Zinc Oxide Based Composite Materials for Advanced Supercapacitors. ChemistrySelect **3**(2), 550–565 (2018)
17. Fang, L., Wua, W., Huang, X., et al.: Hydrangea-like zinc oxide superstructures for ferroelectric polymer composites with high thermal conductivity and high dielectric constant. Compos. Sci. Technol. **107**, 67–74 (2015)
18. Shin, S., Park, D., Jung, J., et al.: Ferroelectric Zinc Oxide Nanowire Embedded Flexible Sensor for Motion and Temperature Sensing. ACS Appl. Mater. Interfaces **9**(11), 9233–9238 (2017)
19. Fangsuwannarak, T., Laohawiroj, S., Rattanawichai, P., et al.: Silicon dots films deposited by spin-coating as a generated carrier addition layer of third generation photovoltaics. Prog. Nat. Sci.: Mater. Int. **31**(2), 192–200 (2021)
20. Fangsuwannarak, T., et al.: Bismuth doped ZnO films as anti-reflection coatings for solar cells. 2013 10th Int. Conf. Electr. Eng./Electron., Comput., Telecommun. Inf. Technol. 1–5 (2013)
21. Rattanachan, S.T., Krongarrom, P., Fangsuwannarak, T.: Influence of annealing temperature on characteristics of bismuth doped zinc oxide films. Am. J. Appl. Sci. **10**(11), 1427–1438 (2013)
22. Krongarrom, P., Rattanachan, S.T., Fangsuwannarak, T.: ZnO doped with bismuth in case of in-phase behavior for solar cell application. Eng. J. **16**(3), 59–70 (2012)
23. Laohawiroj, S., Mangkornkaew, A., Maneedaeng, A., et al.: Silicon composite ink for advanced photovoltaic generation prepared by low-cost technique. J. Renew. Energy Smart Grid Technol. **13**(2), 32–40 (2018)
24. Kabir, M.H., Ali, M.M., Kaiyum, M.A., et al.: Effect of annealing temperature on structural morphological and optical properties of spray pyrolized Al-doped ZnO thin films. J. Phys. Commun. **3**(10), 105007 (2019)
25. Mahmood, A., Munir, T., Fakhar-e-Alam, M., et al.: Analyses of structural and electrical properties of aluminium doped ZnO-NPs by experimental and mathematical approaches. J. King Saud University—Science **34**(2), 101796 (2022)
26. Koralli, P., Songül, F.V., George, M., et al.: Comparative studies of undoped/Al-doped/in-doped ZnO transparent conducting oxide thin films in optoelectronic applications. Chemosensors **10**(5), 162 (2022)
27. Dehkordi, H.A., Kamran, D., Ahmad, M., et al.: Thermal post-annealing and gas concentration effect on liquid petroleum gas sensing characteristics of nanocrystalline zinc oxide thin films. J. Mater. Sci. Mater. Electron. **26**(5), 3134–3142 (2015)
28. Hubbard, C.R., Swanson, H.E., Mauer, F.A.: A silicon powder diffraction standard reference material. J. Appl. Crystallogr. **8**(1), 45–48 (1975)
29. Morkoç, H., Özgür, Ü.: Zinc Oxide fundamentals. Materials and Device Technology WILEY-VCH Verlag GmbH & Co. KGaA, Weinheim, Germany (2009)
30. Wolkin, M.V., Jorne, J., Fauchet, P.M., et al.: Electronic states and luminescence in porous silicon quantum dots: the role of oxygen. Phys. Rev. Lett. **82**(1), 197–200 (1999)

31. Khan, M.I., Bhatti, K.A., Qindeel, R., et al.: Characterizations of multilayer ZnO thin films deposited by sol-gel spin coating technique. Results Phys. **7**, 651–655 (2017)
32. Pankove, J.I.: Optical processes in semiconductors. Dover publications, New York, USA (1976)
33. Omran, M., Fabritius, T., Heikkinen, E.P., et al.: Dielectric properties and carbothermic reduction of zinc oxide and zinc ferrite by microwave heating. R. Soc. Open Sci. **4**(9), 170710 (2017)
34. Kaur, D., Bharti, A., Sharma, T., et al.: Dielectric properties of ZnO-based nanocomposites and their potential applications. Int. J. Opt. **2021**, 9950202 (2021)
35. Li, X.D., Chen, T.P., Liu, P., et al.: A study on the evolution of dielectric function of ZnO thin films with decreasing film thickness. J. Appl. Phys. **115**(10), 103512 (2014)
36. Efil, E., Kaymak, N., Seven, E., et al.: Frequency dependent dielectric properties of atomic layer deposition grown zinc-oxide based MIS structure. Physica B **568**, 31–35 (2019)
37. Kasap, S.O.: Principles of electronic materials and devices. McGraw-Hill, New York (2006)

Assessment of the Glasshouse Application for the Glory Lily Flower in an Indian Climate

Prafull Hishikar, Navneet Parihar, Abhishek Tiwari, Akash Verma, Nuthal Apati Sai Sumanth, Prayag Vishwakarma, Vinod Meena, and Vivek Kumar Gaba(✉)

Department of Mechanical Engineering, National Institute of Technology Raipur, Raipur, India
vgaba.mech@nitrr.ac.in

Abstract. The present analysis describes the study on maintaining the temperature of a greenhouse using a fan system for Glory lily flower in Indian climate. Glory lily has various potential health benefits, including anti-inflammatory, antioxidant, and anti-tumour properties, which make it an important plant in traditional medicine. The study includes various factors depicting internal temperature, vapor pressure, discharge rates, power consumption, and average monthly power consumption for summer and winter seasons. The relationship between temperature, discharge rates, and power consumption, showed that power consumption increases with increasing temperature which peaks during the summer months requiring higher rate of fan discharge. A MATLAB (R2015a) program has been developed to optimize the power usage by predicting the time and duration of its requirement by providing valuable insights into optimizing power usage in greenhouse systems to maintain the desired temperature range for specific crops. Overall, the study provides valuable insights into the energy consumption for maintaining a constant temperature in a greenhouse using a fan system, offering a valuable resource for greenhouse management and optimization.

Keywords: Green house · temperature · vapour pressure · power consumption

1 Introduction

Northern Europe was the origin of the cultivation of exotic plants as early as the Middle Ages, and these plants had to spend the winter in cellars or dark barns that were kept frost-free. When glass became a common material, the lack of light could be lessened by adding windows to one vertical wall, but it was still the biggest flaw in this type of shelter. Any time an adaptation is made, the primary concern is always: How will it affect the greenhouse climate and, in turn, the crop yield and quality. Due to the complexity and dynamic nature of the greenhouse effect, predictions about its effects based solely on empirical data are often inaccurate and even lead to the wrong conclusions. The greenhouse climate conditions can be predicted as a function of greenhouse construction, outside weather conditions, and the attained energy savings if the physical processes involved are properly defined and quantified.

J. Baeyens et al. (Eds.): ESRE 2023, pp. 152–163, 2025.
https://doi.org/10.1007/978-981-96-4345-5_12

According to Kumar et al. [1], high temperatures, significant rainfall, and high humidity are just a few of the unfavourable climatic features that make it challenging to grow crops year-round in India's sub humid subtropics. In subtropical India, greenhouse flower production is growing in popularity as a business due to its aesthetic and financial appeal. However, Pek et al. [2] found that the summer's high temperatures have a detrimental effect on the quantity and quality of flowers. In order to achieve desired results, it is necessary in such areas to lower or regulate greenhouse air temperature so that it is more similar to the surrounding environment. Micro-climate forecasting in greenhouses helps growers and designers manage crop production while also enhancing ventilation and actuating systems. The micro-climate variation in a greenhouse can be investigated using field experiments and simulation methods. As a result, it is essential to create a reliable computer model for simulating the micro-climate of a greenhouse while accounting for the effects of geometry and the kind of ventilation that is suitable for the local agroclimatic condition. In the past, Takakura et al. [3] and Chandra et al. [4] developed a dynamic simulation model that was used to predict the micro-climate of a greenhouse for the climatic conditions of the USA and UK. The micro-climate of an unheated greenhouse was created for China's climatic conditions by Zhang et al. [5] and Shuzhen et al. [6]. The model studies provided a precise description of the experimental values. Another model was developed by Wang and Boulard [7] by including the crop's stomatal resistance parameter as well as natural ventilation. The results of the model using the measured values were in good agreement with little root mean square error. For the climatic conditions in Japan, Sase et al. [8] created a method for forecasting the micro-climate under open roof greenhouses. The results of the ventilation rate model and those that were actually observed agreed well, with an R2 value of 0.96. By enhancing the ventilation rate and cover characteristics, Impron et al. [9] developed a simple climate model for greenhouses that predicts the micro-climate of the greenhouse under Indonesian climate conditions. By building smaller versions of each model to take into account the effects of transpiration and ventilation, the internal microclimate of the greenhouse was predicted. The model results offered a fantastic justification for the experimental findings at 5% error levels. Ganguly and Ghosh [10] developed a thermal model to predict the micro-climate for an Indian greenhouse for growing flowers with natural ventilation without fully accounting for the crop aspects. Additionally, computational fluid dynamics (CFD) served as the basis for the design by using the outside climate as a static boundary condition. Other than these, no models that consider floricultural crops have been created for the Indian subcontinent.

Furthermore, there is not much information available regarding greenhouse work for floricultural crops in India. The effects of crop transpiration on temperature and vapour pressure, as well as the effects of natural ventilation on the sides and roof of the structure, were taken into consideration to create the current greenhouse climate model for India's favourable conditions for floricultural production. The results of field tests carried out with a Gerbera crop in the Sawtooth greenhouse were used to calibrate and validate the developed model. Sawtooth is a type of structure in which two or more frame are connected to one another along its length. This type of structure provides natural ventilation for Indian climatic condition. The results of field tests carried out with a Glory lily in the Sawtooth greenhouse were used to calibrate and validate the developed model. Glory lily has been used in traditional medicine for centuries due to its many health benefits. Among other potential health benefits, Glory lily has anti-inflammatory, antioxidant, and anti-tumor properties that may lessen the signs and symptoms of conditions like arthritis and lower inflammatory levels in the body. Additionally, it might protect cells from free radical harm and aid in halting the development and spread of cancer cells within the body as per the reports of Traditional Medicine Research (TMR) journals [11].

2 Numerical Method and Its Validation

2.1 Model Development

The following assumptions were made when developing a greenhouse climate model:

(i) Greenhouse was thought of as having a good mix of compartments [12].
(ii) The cover's and shade net's heat storage capacities were insignificant in comparison to the greenhouse's current fluxes [13].
(iii) It has been determined that an interior shading screen has little to no effect on thermal radiation reflectivity.
(iv) Because the model was created to simulate the dynamics of the ambient climate under static conditions, it was assumed that the soil heat storage flux in the greenhouse would be zero over the course of a 24-h period.
(v) The boundary conditions included air temperature, the strength and direction of the wind from the outside, and solar radiation. The greenhouse can be represented by a simple linear balance.

$$\mu Go - Ks\Delta T - K_L\Delta e - K_c\Delta T - Q_m = 0 \tag{1}$$

where G_o is the external radiation (W/m^2) which falls on the greenhouse structure, and $\Delta T = Ti - To$ is the temperature difference between the inside and outside air (K), Ks is the sensible heat transfer coefficient (W/m^{2}K), K_L is the latent heat transfer coefficient, and Kc is the overall heat transfer coefficient, μ is solar heating efficiency and $\Delta e = ei - eo$ is the water vapour pressure difference between the inside and outside air (Pa).

2.2 Ventilation Submodel

According to Boulard et al. [14], aspect ratio (F_o), is the function of length and width of greenhouse structure which helps in regulating discharge through the structure. A single rectangular opening's air flow resistance was written as follows:

$$F_o = 1.75 + 0.7exp\left[\frac{-1/w}{32.5sin\alpha}\right]$$
$$C_d = F_o^{-0.5} \tag{2}$$

where C_d is the discharge coefficient, which is helpful in figuring out the flow circumstances in the phenomenon of natural ventilation, and the ventilation rate (m^3/s) is expressed as

$$Q = \frac{A}{2} C_d \left(2g \frac{\Delta T}{T_0} \frac{w}{2} + C_w U_e^2 \right)^{0.5} \tag{3}$$

$$Q = C_d \left(2g \frac{\Delta T}{T_Q} \left(\frac{(A_R A_S)^2}{A_R^2 + A_S^2} \right) + \left(\frac{A_R + A_S}{2} \right)^2 C_w U_0^2 \right)^{0.5} \tag{4}$$

where U_o is the outside wind speed at 4 m height, A is the ventilation area (m^2), and C_w is the wind pressure coefficient.

2.3 Transpiration Submodel

From the Pennman–Monteith formula, stomatal resistance and the aerodynamic resistance, whereas stomatal resistance is offered by the plant stem in transferring stored water from its roots to leaves also known as transpiration, while aerodynamic resistance emphasises on the boundary layer formed on the plant canopy, greenhouse transpiration can be calculated as follows:

$$\lambda E = \frac{\rho C_p l_{LA}}{\gamma} \left(\frac{e_f^* - e_C}{r_s + r_\alpha} \right) \tag{5}$$

The following equation was used to determine the aerodynamic resistance, r_a

$$r_a = 305 \left(\frac{d_v}{U_i} \right)^{0.5} \tag{6}$$

Boulard et al. [14] developed a straightforward empirical relationship from which the stomatal resistance, r_s, was derived.

$$r_s = 200 \left(1 + \frac{1}{\exp(0.05(\mu G_0 - 50))} \right) \tag{7}$$

As a result, at leaf temperature, the pressure of saturated water vapour is different from the air's interior water vapour pressure.

$$e_f^* - e_i = \delta(T_i) \Delta T_f + D_i \tag{8}$$

The value of ΔT_f is represented as and expressed in relation to the difference in air temperature between the inside and outside of the greenhouse.

$$\Delta T_f = \Delta T_{fo} - \Delta T \tag{9}$$

It is possible to express the difference in the internal air's water vapour pressure (D_i) as a linear function of the temperature and ambient air water vapour pressure deficit. Do and T_o

$$D_i = \left(e^*(T_i) - e^*(T_0) \right) - (e_i - e_0) + \left(e^*(T_0) - e_0 \right) \tag{10}$$

where e_o is the water vapour pressure of the outside air (Pa) and e* (T) is the saturated water vapour pressure of air at temperature T. If there is ventilation present and the inside temperature (T_i) is the same as the outside temperature (T_o), then D_i can be roughly calculated as

$$D_i = \delta(T_O)\Delta T - \Delta e + D_O \tag{11}$$

Thus, combining Eqs. (9), (10) and (11) yields

$$e_f^* - e_i = \delta(T_i)\Delta T_{fo} - (\delta(T_i) - (T_o))\Delta T - \Delta e + D_o \tag{12}$$

The greenhouse crop transpiration E can also be written as if the capacitive term in the greenhouse's water vapour balance were ignored.

$$\lambda E = K_L \Delta e \tag{13}$$

It is possible to infer the following relationship from Eqs. (6), (12), and (13):

$$\left[\frac{\gamma(r_s + r_a)}{\rho C_p L_{LA}} K_L + 1\right]\Delta e = \delta(T_i)\Delta T_{fo} - (\delta(T_i) - (T_o))\Delta T + D_o \tag{14}$$

Finally, a linear system of three equations with three unknowns (ΔT, ΔT_{fo}, and Δe) were developed from Eqs. (1)–(14) as

$$(K_s + K_c)\Delta T + K_L \Delta e = \mu G_O - Q_m \tag{15}$$

A linear system of three equations with three unknowns was subsequently developed from Equations.

$$(K_s + K_c)\Delta T + K_L \Delta e = \mu G_o - Q_m \tag{16}$$

The unknown variables ΔT, ΔT_{fo}, and Δe can be analytically determined given the characteristic parameters for the greenhouse (Kc and μ), the crop (Leaf Area Index, r_a, and r_s), and the measured boundary conditions for the outside climate (G_o, T_o, and D_o).

2.4 Parameters Estimation

Estimation of the Heat Transfer Coefficients (K_s and K_L) for Sensible and Latent Heat. The aforementioned equations can be used to deduce that the sensible and latent heat transfer coefficients between the greenhouse's interior and exterior are proportional to the ventilation flux.

$$K_s = \frac{\rho C_p Q}{A_g} \tag{17}$$

$$K_L = \frac{\lambda \zeta \rho Q}{A_g} \tag{18}$$

The correlation was used to estimate the external wind speed (Uo) in m/s 1 and Kc, which determines the overall heat transfer coefficient between the inside and outside of the greenhouse.

$$K_c = A + BU_0 \tag{19}$$

where greenhouse aspect ratio and cover characteristics determine the values of A and B. Measurements of crop transpiration in the greenhouse where the gerbera crop was grown allowed for the identification of the parameter's values. After the measurements for each season were collected, the average values of A and B were calculated and discovered to be 6.5 and 0.4, respectively.

Calculating the Effectiveness of Solar Heating in Greenhouses (μ). The amount of solar energy used during the day to heat the interior air, soil, plants, and other structures in greenhouses is referred to as the greenhouse's solar heating efficiency (μ). . Solar heating efficiency is the term used for the portion of heat available to heat the inside environment. Additionally, it influences the evaporation of plant substrate, soil water, and transpiration. It was a crucial parameter in the model's development because it takes greenhouse design and geometry into account as well as the variation in soil heat flux. The equation was used to determine the solar heating efficiency by taking into account the fact that the total soil heat storage (Q_m) was nearly zero over a 24-h period.

$$\mu = \frac{\sum l(K_s + K_c)\Delta T + K_L \Delta e]}{\sum G_o} \tag{20}$$

The micro-climate of the greenhouse was predicted using a computer program written in the MATLAB (R2015a). Crop data gathered during the winter and summer growing seasons of the year, as well as measured internal and external climate data for the greenhouse, were all taken into consideration as inputs for the computer software. To calculate the interior air temperature, canopy temperature, and interior vapour pressure, the multi variable Newton-Raphson iteration method is used to solve the system of three linear Eqs. (16)–(18). Multi-variable Newton Raphson method performs iterations on Eqs. (12), (14) and (16) and flowchart were used to calculate these three state variables. Table 1 represents the Variables used in greenhouse climate model.

3 Result and Discussion

The present analysis describes the study on maintaining a greenhouse's temperature using a fan system. There are several graphs in the study, including ones showing internal temperature, vapour pressure, discharge rates, and power consumption for the summer and winter. Overall, the study offers insightful information about the amount of energy required to use a fan system to maintain a constant temperature in a greenhouse. The validation of the present analysis is shown in Fig. 1a, where the temperature measured with a thermometer and the inner temperature obtained through MATLAB (R2015a) code are represented on the x-axis and y-axis, respectively. The graph demonstrates a good agreement between the current result and Kumar *et al.* [1]. Another validation for the present methodology has been shown in Fig. 1b with respect to vapour pressure, which also shows good agreement with Kumar et al. [1].

Table 1. Variables used in greenhouse climate model

Variables	Values
A_g– Area of Greenhouse (m^2)	14x6
A_R– Area of Roof Opening (m^2)	14x0.35
A_S– Area of Side Opening (m^2)	14x0.35
G_0– Global Solar Radiation (W/m^2)	Measured using SAM software
c_p– Specific heat of Air (J/KgK)	1005
T_0– Exterior Air Temperature (K)	Measured using SAM software
e_0– Exterior Air Vapour Pressure (Pa)	Measured using SAM software
r_s– Stomatal Resistance (s/m)	200
r_a– Aerodynamic Resistance (s/m)	96.44
γ– Psychometric Constant (Pa/K)	0.000665xP where P is the atmospheric pressure
K_L– Latent heat transfer coefficient $\left(W/m^2Pa\right)$	61.797
I_{LA}– Leaf area index expressed as leaf area per unit ground area	0.0333
δ– Slope of the saturation air water vapour (Pa/K)	0.43071
K_C– Overall heat transfer coefficient between inside and outside $\left(W/m^2K\right)$	7.1
K_S– Sensible heat transfer coefficient $\left(W/m^2K\right)$	439.687
Q_m– Soil heat flux (W/m^2)	0 (Net heat flux within a Day)
Q– Air flow $\left(m^3/s\right)$	
T_i– Interior Air Temperature (K)	Measured using SAM software
T_{can}– Plant Canopy Temperature (K)	Initial Guess (Output By Iteration)

In Fig. 2, the fan's discharge rates for Summer and Winter are shown (in metre cubes per second). The results display the fan discharge that needs to be kept at 25 °C, 30 °C, and 35 °C for Summer and 25 °C and 30 °C for Winter in the greenhouse. The fan discharge must increase to maintain the greenhouse's temperature because the discharge rate is increasing as the temperature does. The cause of the evening drop in discharge is the same. All line graphs reach their peak around the 13th hour due to the day's highest temperature.

Figure 3 shows the power usage (in watts) for the summer and winter seasons. The graphs display the amount of energy used by the fan to maintain the greenhouse's temperature at 25 °C, 30 °C, and 35 °C, respectively. These graphs show that the summer power requirement values are more than twice as high as the winter values. The power

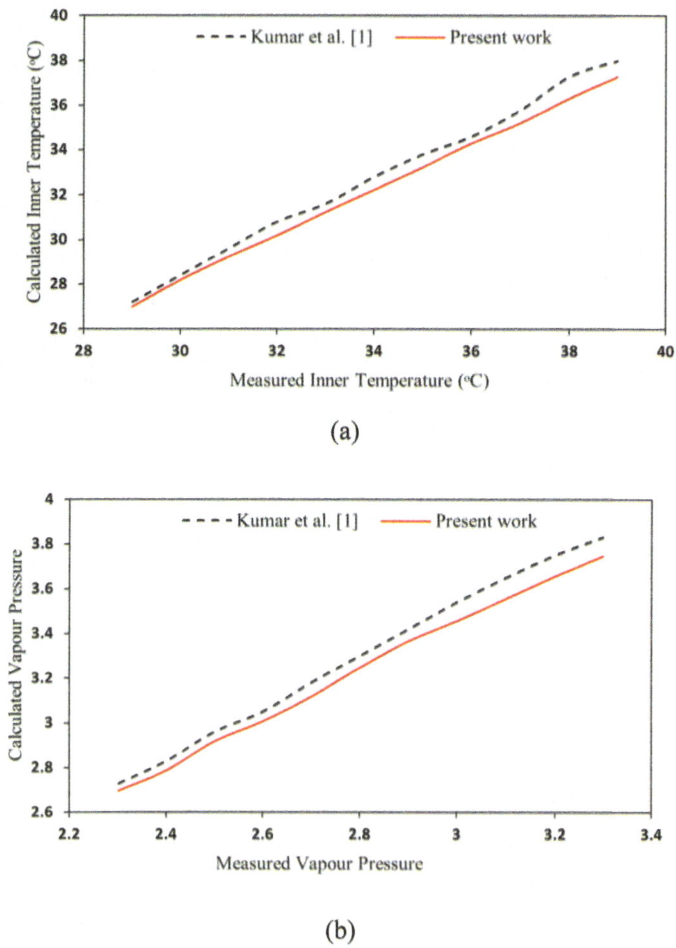

Fig. 1. Variation of (a) inner temperature (b) vapour pressure

usage for the summer has been seen to start at a minimum of 150 watts around 6 a.m., peak at 650 watts around the 14-h mark, and then gradually decline until the 20-h mark. The fan discharge should rise with the temperature to maintain the greenhouse at a constant temperature, so the power consumption has increased since the morning. As a result, afternoon is when the most power is needed. The same factor is responsible for the evening's lower energy use.

Figure 4 shows the average monthly power consumption (watt). This graph displays the energy required by the fan to discharge air and keep the greenhouse at a temperature of 30 °C. The average monthly power consumption rises throughout the year, reaching its peak in the summer because the fan discharge must rise to maintain the green house's temperature (April to June). However, as summer came to an end, power usage started to decrease.

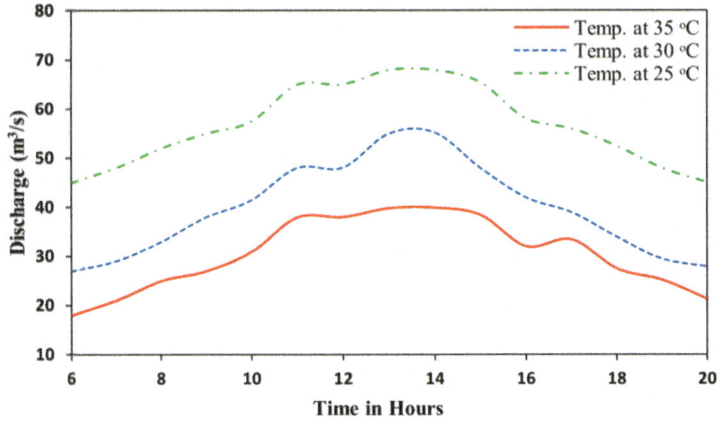

(a)

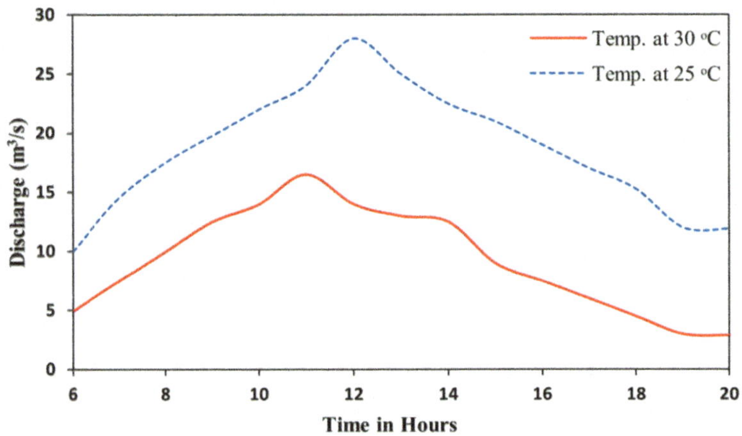

(b)

Fig. 2. Fan discharge analysis for (a) Summer season and (b) Winter season

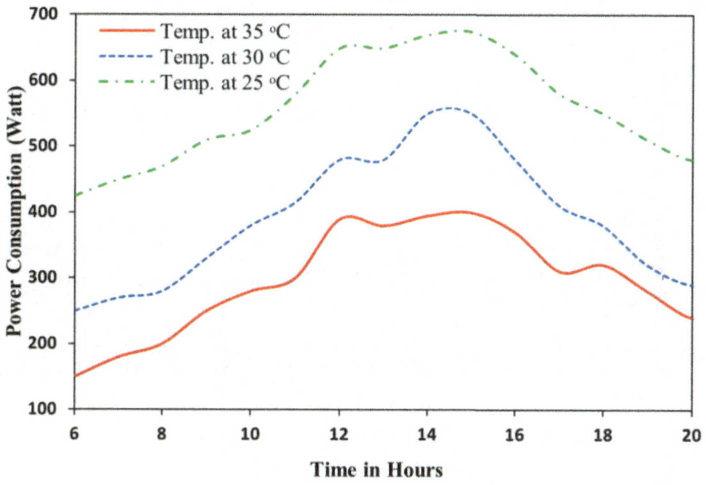

(a)

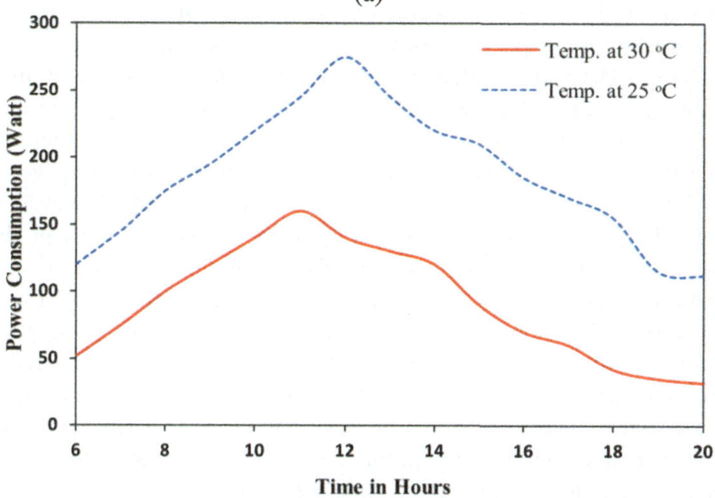

(b)

Fig. 3. Fan power consumption for (a) Summer season and (b) Winter season

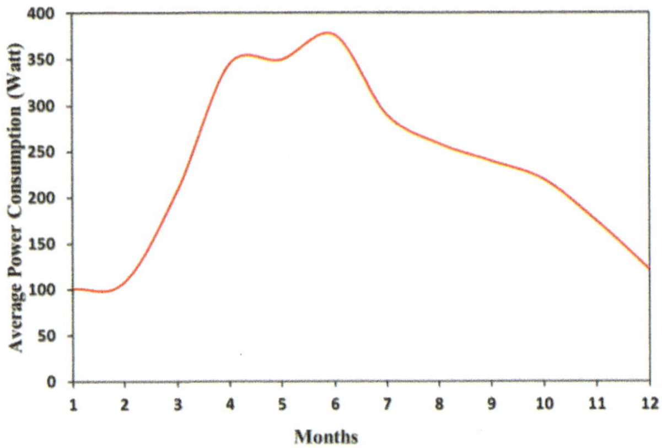

Fig. 4. Average monthly fan power consumption

4 Conclusion

The present study suggests that discharge plays a critical role in maintaining the temperature of a greenhouse, especially for the glory lily crop that requires a temperature range of 30–35 °C. The main challenges in the cultivation of glory lily are maintaining the desired environment with an economical approach. Optimizing inner atmosphere needs and providing sufficient ventilation can helps in achieving this target. It has been observed that fluctuating the discharge rates can help maintain the required temperature, but it also increases the power requirement for the system. However, the study presents a MATLAB (R2015a) program that optimizes power usage by predicting the time and duration of the power requirement. This program also helps us in selecting suitable location for the cultivation of medicinal plants like Glory lily. This program can predict the inner microenvironment of the greenhouse and ensure that the required temperature range is maintained efficiently. Methods used in obtaining desired programs are Multi-variable Newton Raphson iteration and finite difference methods. Initial guess has been made and by using these two methods iterations is performed. Overall, the study provides valuable insights into optimizing power usage in greenhouse systems to maintain the desired temperature range for specific crops.

In present work problems related to optimization of environment needs and achieving desired condition by predicting environment has been done. This work requires continuous monitoring for its efficient working hence automation of all systems is required to minimize the cost related to labour and also avoid the danger of crop damaging. Hence in future inclusion of IOT devices will help in providing better crop quality and cost effectiveness.

References

1. Kumar, K.S., Rao, K.A., Mittra, B.N.: Low-cost plastic greenhouses-design, development and their suitability for raising seedlings. An unpublished M.Tech. Thesis, Agricultural and Food Engineering, Indian Institute of Technology, Kharagpur (1991)
2. Pek, Z., Hayles, L.: The effect of daily temperature on truss flowering rate of ornamental crops. J. Sci. Food Agric. **84**(13), 1671–1674 (2004)
3. Takakura, T., Jordan, T.A., Boyd, L.L.: Dynamic simulation of plant growth and environment in the greenhouse. Trans. Am. Soc. Agric. Eng. **14**(5), 964–971 (1999)
4. Chandra, P., Albright, L.D., Scott, N.R.: A time dependent model of the greenhouse thermal environment. Trans. Am. Soc. Agric. Eng. **24**(2), 442–449 (1981)
5. Zhang, Y., Mahrer, Y., Margolin, M.: Predicting the microclimate inside a greenhouse: an application of a one-dimensional numerical model in an unheated greenhouse. Agric. For. Meteorol. **86**(3–4), 291–297 (1997)
6. Shu-zhen, L., Young, H.E., Zhang, Y., Xiang-wen, M.: Prediction and analysis model of temperature and its application to natural ventilation multi span greenhouse equipped with insect proof screen. J. Zhejiang Univ. Sci. B **6**, 523–529 (2005)
7. Wang, S., Boulard, T.: Predicting the microclimate in a naturally ventilated plastic house in a Mediterranean climate. J. Agric. Eng. Res. **75**(1), 27–38 (2000)
8. Sase, S., Reiss, E., Both, A.J., Roberts, W.J.: Developing a natural ventilation model for open-roof greenhouses, accessed through internet. http://www.cook.rutgers.edu/~horteng/ppt/papers/OpenRoofModel.pdf
9. Impron, I., Hemming, S., Bot, G.P.A.: Simple greenhouse climate model as a design tool for greenhouses in tropical low land. Biosys. Eng. **98**(1), 79–89 (2007)
10. Ganguly, A., Ghosh, S.: Model development and experimental validation of floriculture greenhouse under natural ventilation. Energy Build. **41**(5), 521–527 (2009)
11. Traditional Medicine Research (TMR) journals (ISSN 2413-3973)
12. Roy, J.C., Boulard, T., Kittas, C., Wang, S.: Convective and ventilation transfer in greenhouses. Part 1: the greenhouse considered as a perfectly stirred tank. Biosyst. Eng. **83**(1), 1–20 (2002)
13. Bot, G.P.A.: Greenhouse climate: from physical processes to a dynamic model, Ph.D. Thesis, Wageningen Agricultural University, the Netherlands (1983)
14. Boulard, T., Meneses, J.F., Mermier, M., Papadakis, G.: The mechanisms involved in the natural ventilation of greenhouses. Agric. For. Meteorol. **79**(1–2), 61–77 (1996)
15. Kindelan, M.: Dynamic modeling of greenhouse environment. Trans. ASAE 1232–1239 (1980)

Reviewing the Production of Fine Particles and Powders

Miao Yang[1], Shuo Li[1], Huili Zhang[2], Zehao Li[2], Xinyu Dong[3], Jan Baeyens[1], and Yimin Deng[4(✉)]

[1] Beijing Advanced Innovation Centre for Soft Matter Science and Engineering, Beijing University of Chemical Technology, Beijing 100029, China
[2] College of Life Science and Technology, Beijing University of Chemical Technology, Beijing 100029, China
[3] College of Mathematics and Science, Beijing University of Chemical Technology, Beijing 100029, China
[4] School of Chemistry and Chemical Engineering, Beijing Institute of Technology, Beijing 100081, China
dengyimin0718@163.com

Abstract. Fine and ultrafine particles are of growing importance for numerous industrial processes. Some examples are the use in the cement and sand industry, the production of explosives and pharmaceuticals, ore refining, the catalyst powders and fertilizers in the chemical industry, extenders in macromolecular compounds (paints, cosmetics, resins), and the use in space and nuclear industry. Ultrafine powders hence have widespread applications, due to their excellent properties: very high chemical activity due to their high specific surface area), low melting and sintering point, high specific heat among others. Different techniques can produce ultrafine particles. Mechanical production comprises crushing and grinding of solids, which remain the main techniques. There are however several other techniques available of so-called physic-chemical nature: thermal decomposition of solids, deposition from liquids, vapour phase methods and solid-state reactions. The energy consumption and obtained quality are often the selection criteria.

Keywords: Fine particles · Milling · Particle size · Energy

1 Mechanical Production of Fine Particles

Size reduction of solids increases the surface area for reaction and thereby increases the chemical reactivity, liberates particles for recovery and results in better mixing characteristics. Solids are comminuted by applying force. Different methods of applying force are: (i) impact through particle concussion by a single rigid body; (ii) compression leading to disintegration by two rigid bodies; (iii) shear through particle interaction produced by fluid or particle; and (iv) attrition by rubbing against each other or against a rigid surface. Compressive actions result in rather large (100 μm), impact in intermediate (10 μm) and shear in small particle sizes (1 μm). Different crushers and grinders use different methods of applying force, as illustrated in Table 1.

J. Baeyens et al. (Eds.): ESRE 2023, pp. 164–180, 2025.
https://doi.org/10.1007/978-981-96-4345-5_13

Table 1. Applied forces in different comminution equipment

comminution equipment	compression	impact	friction	shear
jaw crusher	x			
gyratory crusher	X			
roll crusher	X			x
edge runner	X		x	x
hammer mill		X		
ball mill		X	x	
jet mill		X	X	
disc crusher			x	

The milling process is energy intensive and highly inefficient and often quoted as being responsible for 5% of the countries' power generation. Compared to the energy requirement for the creation of a new surface, the efficiency of the industrial scale process is generally less than 1%. Despite considerable research over the past years, size reduction processes today remain stubbornly inefficient. Moreover, despite a well-developed theory for the strength and breakage mechanisms of solids, design and scale-up of comminution processes are usually based on experience and tests, and depend greatly on the manufacturer of the size reduction equipment [1, 2].

1.1 Particle Fracture Mechanisms

As a simple and convenient model of a brittle material, a crystal of sodium chloride (salt) is considered. Such a crystal consists of a lattice of positively charged sodium ions and negatively charged chloride ions, arranged in such a way that each ion is surrounded by six ions of the opposite sign: the lattice is shown in Fig. 1.

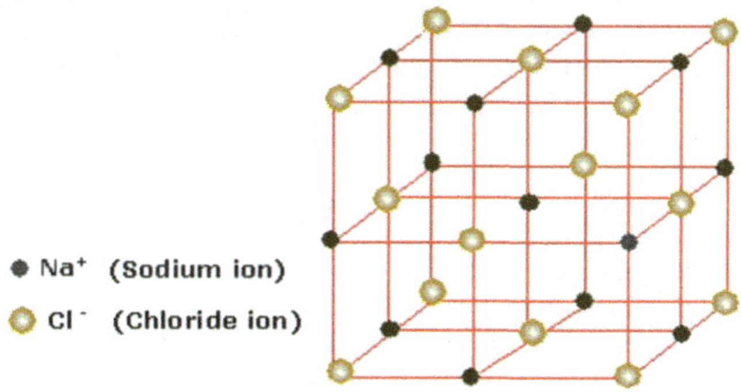

Fig. 1. Sodium chloride crystal lattice [3]

The oppositely charged ions are attracted by a force inversely proportional to the square distance between the ions. At very small interatomic distances, the repulsive force between the negatively charged electron clouds becomes important. These two forces result in an equilibrium separation between two oppositely charged ions, making repulsive and attractive forces equal though opposite. The sum of the attractive and repulsive forces varies with the interatomic distance, as shown in Fig. 2.

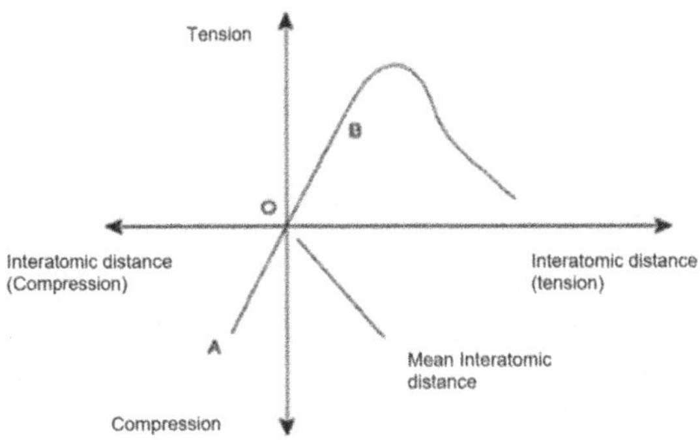

Fig. 2. Force versus distance on an atomic scale: F = force between atoms; dia = interatomic distance

Within a small range of interatomic distances, the relationship between the applied compressive or tensile force and the resulting change in ion separation is linear, indicated by AB in Fig. 2, Hooke's law applies as the strain is directly proportional to the applied stress. To break the crystal, adjacent layers have to be separated beyond the region where Hooke's law applies, i.e. beyond point B in Fig. 2. when the repulsive and attractive forces in a crystal are known, the strength of a crystal can hence be estimated.

A body submitted to tension stores energy, the so-called strain energy. When stress is plotted versus strain, the area under the graph is the amount of the strain energy stored in a brittle material under tension. This strain energy is however not uniformly distributed throughout the body, but concentrated around cracks and holes. To account for these cracks and holes, Inglis (1913) introduced the stress concentration factor K, calculated according to:

$$K = 1 + 2\sqrt{\frac{L}{R}} \tag{1}$$

with L = half the length of the crack

R = radius of the crack tip or hole

For a crack to propagate, Griffith [4] specifies that the following criteria must be met: (i) the strain energy released must exceed the surface energy created and (ii) there must be a crack propagation mechanism available. Furthermore, for a given mean stress applied to a body, there should be a critical minimum crack length where the stress concentration at the tip will be just sufficient to cause crack propagation. Due to this stress, a crack initially longer than the critical length will grow longer until the body is broken. During crack growth at a constant stress, there is strain energy in excess of that required for crack propagation. This excess energy is dissipated in the material at the velocity of sound: it concentrates at the tips of other cracks, causing them to propagate as well. The rate of crack propagation is lower than the velocity of sounds, which results in the propagation of other cracks before the first crack brings about failure: thus, brittle materials will have multiple fractures. For materials like rubber, plastics and metals, excess strain energy rather causes deformation of the material [1].

Small particles are more difficult to break than large particles. This can be explained by using the concept of failure by crack propagation. Firstly, the length of a crack is limited by the particle size, so lower maximum stress concentration factors can be achieved in small particles. Lower stress concentrations result in higher mean stresses to cause failure. Secondly, for small particles, the equation of Inglis overestimates the value of K, as there is less room for the development of stress patterns. The maximum stress concentration is hereby limited, so a higher mean stress is necessary to cause crack propagation.

1.2 Energy Requirement and Product Size Distribution

Energy Requirements. Several formulas have been proposed to predict the energy required for particle size reduction. In 1867 Rittinger proposed that the energy requirement is directly proportional to the area of new surface created. In practice however, the energy requirement is usually 200–300 times that required for the creation of a new surface: the formula proposed by Rittinger is unlikely to represent the energy required. In 1885 Kick proposed that the energy required is directly proportional to the ratio of the particle volume before versus the particle volume after comminution. This statement would implicate that the same energy is required to reduce 10 μm particles to 1 μm particles as to reduce 1 m rocks to 10 cm blocks. This is clearly not true, which makes the proposed formula unrealistic. A more useful formula – for products with top size below 100 μm - was suggested by Bond in 1952, based on data obtained from industrial scale processes involving many materials:

$$E = W_I\left(\frac{10}{\sqrt{x_2}} - \frac{10}{\sqrt{x_1}}\right) \tag{2}$$

with E = the energy required to reduce the top particle size from X_1 to X_2 [kWh/short ton]

W_I = the Bond work index = the energy required to reduce the size of a unit mass of the material from infinity to 100 μm in size [kWh/short ton = 4000 J/kg].

Since the top particle size is difficult to define, X_1 and X_2 are taken to be the sieve size in μm through which 80% of the material will pass. The Bond work index must be determined empirically, and values are given in [5].

In 1957, *Holmes* proposed a differential equation with Rittinger, Kick and Bond being special cases.

$$\frac{dE}{dX} = -C\frac{1}{X^N} \tag{3}$$

with N = 2 for Rittinger =>

$$E = C_R\left(\frac{1}{L_2} - \frac{1}{L_1}\right) \tag{4}$$

N = 1 for Kick =>

$$E = C_K \ln\left(\frac{L_1}{L_2}\right) \tag{5}$$

N = 1.5 for Bond =>

$$E = C_B\left(\frac{1}{\sqrt{L_2}} - \frac{1}{\sqrt{L_1}}\right) \tag{6}$$

These different approaches provide fair results within certain areas of product size, as shown in Fig. 3. Kick's proposal is applicable for large particle sizes, Rittinger for very small particle sizes, and the Bond formula is suitable for intermediate sizes, being the most common range for comminution processes. It is however advisable to rely on the experience of equipment manufacturers and on test results to predict the energy requirement for milling particular materials.

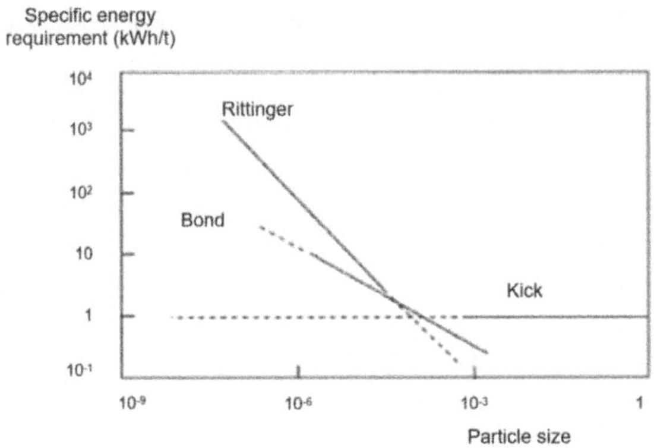

Fig. 3. Comparison of different proposals predicting the energy requirement for particle size reduction

Product Size Distribution. The milling process in comminution equipment is modelled based on both the specific rate of breakage, and the breakage distribution function. The specific rate of breakage S(x) is the probability of a particle of size x being broken in a unit time. The breakage distribution function b(z,x) determines the fraction of breakage product from size interval x into size interval z. The meaning of these two functions is demonstrated in Fig. 4 for a 10 kg sample of monosized particles with x = 1. If S(1) = 0.6, 4 kg of material will remain in size interval 1 after unit time, with a size distribution of the breakage product described by a set of b(z,x) values. If for example b(4,1) = 0.25, 25% of the material from size interval 1 will fall into size interval 4. The breakage distribution function can also be expressed cumulatively as B(z,x), which means that from a particle size x a proportion B(z,x) of the product falls below size z [6].

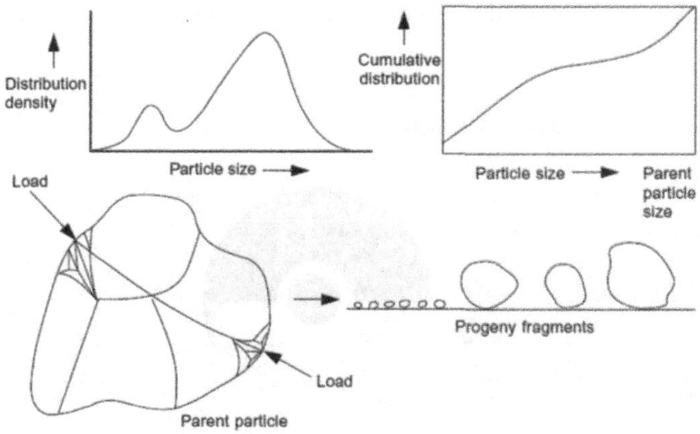

Fig. 4. Meaning of specific rate of breakage and breakage distribution function [6]

Using this concept, the breakage product can be characterized as follows. The mass of breakage product of size <z is calculated by multiplying the mass of feed by S(x) and by B(z,x). The mass of breakage product of size >z equals the mass of feed multiplied by S(x) and by 1 – B(z,x). The mass of unbroken feed in the product is determined by multiplying the mass of feed by 1 – S(x). This example is based on a monosized feed with a breakage function that splits the product in only two fractions. In practice, a more realistic size distribution and sophisticated breakage functions will occur: the same method however can be used, by simply replacing the feed size, selection function and breakage function by matrices.

2 Comminution Equipment

For a particular grinding equipment, the choice of equipment depends on several variables: stressing mechanisms, particle size of feed and product, material properties, carrier medium, mode of operation, capacity, and the combination with other unit operations.

2.1 Stressing Mechanism

Particle size reduction in mills can be caused by three stressing mechanisms, which can be used for initial the classification of comminution equipment:

- Crushing + attrition: stress applied between two surfaces (surface - particle, or particle - particle) at low velocities (0.01–10 m/s);
- Impact fracture + attrition: stress applied on a single solid surface (surface - particle, or particle - particle) at high velocities (10–200 m/s);
- Disagglomeration: stress applied by a carrier medium (usually in wet grinding).

2.2 Crushing Equipment (Stressing Mechanism 1)

Figure 5 shows a jaw crusher, which behaves like a pair of giant nutcrackers. One jaw is fixed, the other can move by means of toggles driven by an eccentric. The material, mostly of rock-size, is crushed between the jaws and the product leaves the crusher at the bottom.

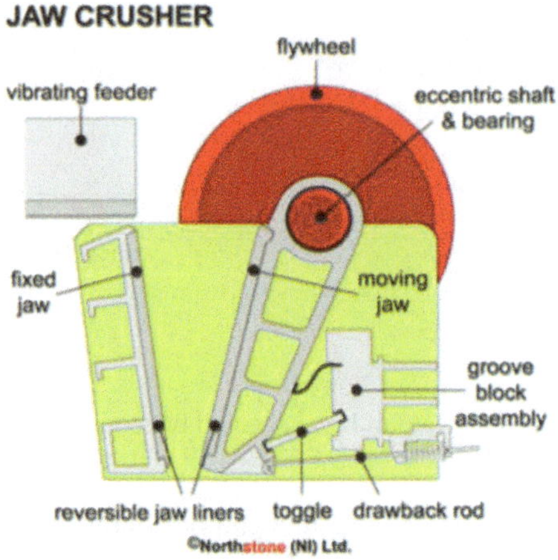

Fig. 5 Jaw crusher [7]

The gyratory crusher of Fig. 6, consists of a truncated cone (the fixed jaw) and an eccentrically rotating cone (the moving jaw). When the solids are fine enough, they are discharged through the gap between the jaws.

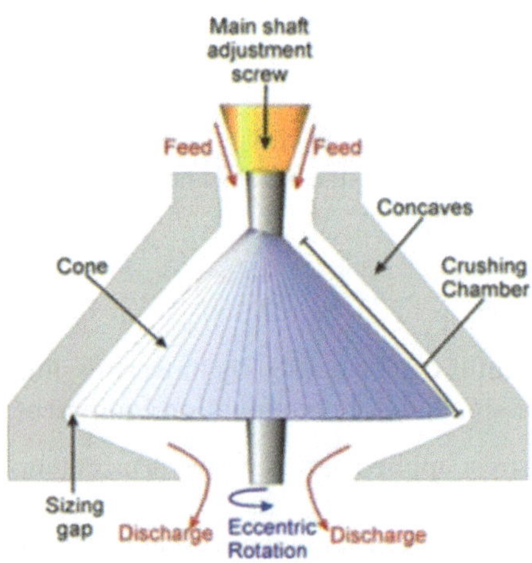

Fig. 6. Gyratory crusher [8]

Figure 7 shows a rolls crusher: two cylindrical rolls rotate in opposite directions, horizontally and side by side with an adjustable gap. Material is dragged in by the rotation of the rolls, and particles are broken when the passing through the gap.

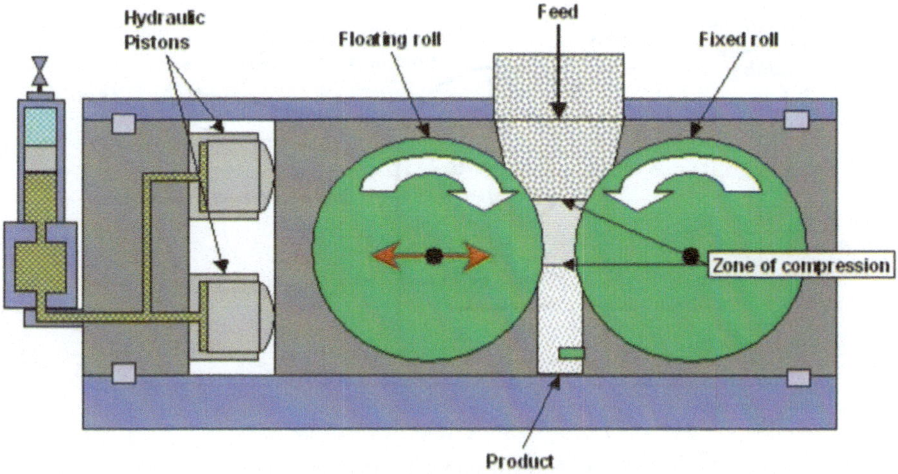

Fig. 7. Crushing roll [9]

Alternatives and additional crushing equipment of similar mechanism include the horizontal table mill, with a circular rotating table and a crushing roller; and the pendular mill, with multiple rotating rollers being swung centrifugally against the wall of the crusher while ploughs feed the feedstock towards the impact surface.

2.3 Equipment with High Velocity Impact (Stressing Mechanism 2)

A hammer mill consists of a rotating shaft with fixed or pivoted hammers, rotating inside a cylinder. Material can be fed by gravity or by a gas stream. In the gravity fed version, particles are able to leave the chamber small enough to pass through a grid at the bottom of the mill. When material is fed by a gas stream, a fan is needed at the outlet of the chamber to keep the particles in motion, as shown in Fig. 8. The air flow rate determines the cut-size of the mill.

A pin mill consists of two parallel circular discs carrying a set of projecting pins. Either the first disc is fixed and the other one rotates at high speed, or both rotate in opposite direction. Particles are fed into the centre, they move radially outwards, and are fractured by impact or attrition.

In a fluid energy mill, turbulence is created in high velocity jets of air or steam: this produces conditions for inter-particle collisions that cause particle fractures. A commonly used type of fluid energy milling is the loop or oval jet mill, shown in Fig. 9. The jets are placed at the base of the loop, where the material is grinded. After grinding, the particles are conveyed to the classifier and exit at the top of the loop. Fluid energy mills consume much energy and are not suitable for handling abrasive materials. A fluidized bed jet mill, where multiple jets are positioned along the circumference of the bed, resolves these problems to some extent, as the bed absorbs the energy from the high speed particles.

Equipment using a combination of crushing and impact with attrition (stressing mechanisms 1 and 2).

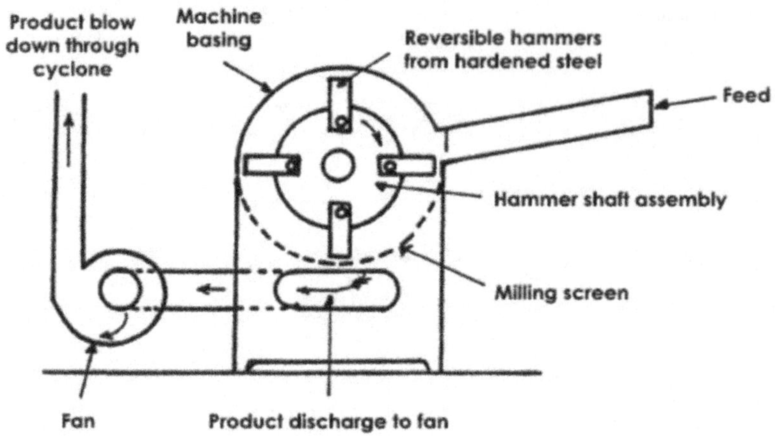

Fig. 8. Gas stream fed hammer mill [1]

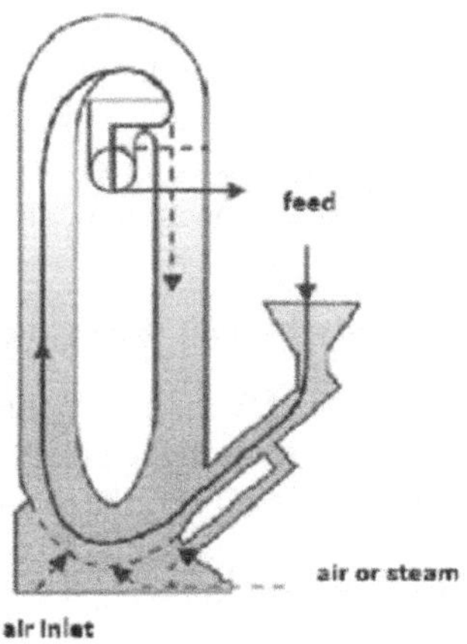

Fig. 9. Loop Jet mill

A sand mill, as shown in Fig. 10, consists of a vertical cylinder containing a stirred bed of sand or glass beads. Particles are fed in the form of slurry into the bottom of the bed, the product passes through the cylinder and is discharged at the top through a screen which retains the bed material.

In a colloid mill, the particles are also fed as slurry. The mill consists of a male ribbed cone rotating at high speed and a female static cone. The slurry passes through the gap between both cones.

Figure 11 shows a ball mill, a rotating cylindrical or conical shell half filled with balls of steel or ceramic. Due to the (strictly controlled) rotational speed of the shell, the balls tumble one over another without causing cascading. Both balls and particulate material are subject to a centrifugal force, and particles are milled in contact with the

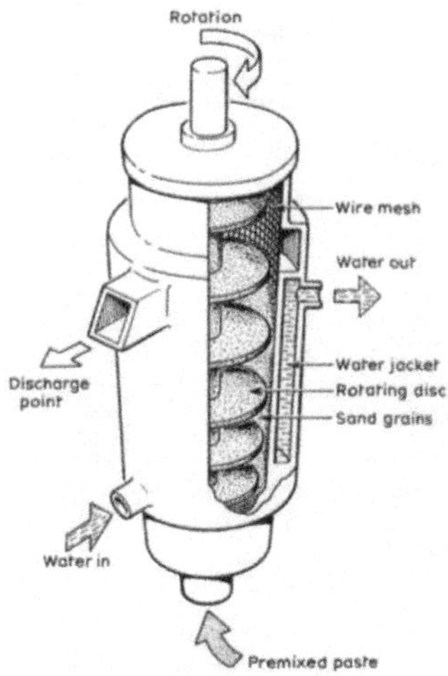

Source: Turner (1998)

Fig. 10. Sand mill [5]

balls and the wall. The critical speed of a ball mill can be calculated by the following formula:

$$v = \frac{42.3}{\sqrt{D - d}} \tag{7}$$

with D = mill diameter [m]
d = ball diameter [m]

2.4 Particle Size

In most cases, the choice of a suitable mill is determined by the size of the feed and the desired product size distribution, as illustrated in Table 2.

2.5 Material Properties

Material properties affect the choice of an appropriate mill, though to a lesser extent than feed and particle size. When selecting a mill, the following characteristics should be considered:

- Hardness as a measure of the resistance to abrasion, usually measured on the Mohs' scale of hardness (graphite = 1 and diamond = 10);
- Abrasiveness, linked closely to hardness and probably the most important factor to select commercial mills;
- Toughness as a measure of the resistance to propagation of cracks: in tough materials excess strain energy will result in plastic deformation rather than in grinding, a problem that can be solved by operating at lower temperature, rendering the material more brittle;

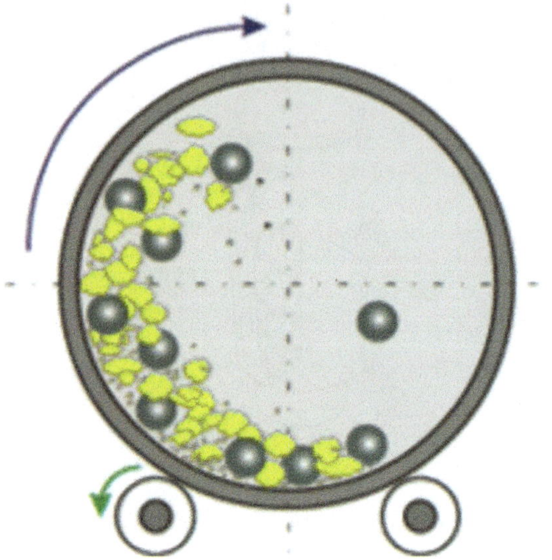

Fig. 11. Ball mill [10]

Table 2. Influence of product size on type of mill

<50 μm	50 μm–3 mm	>3 mm
ball mills vibration mills sand mills perl mills colloid mills fluid energy mills	ball mills rod mills pin mills tube mills vibration mills	crushers table mills edge runner mills

- Cohesivity and adhesivity, the properties whereby particles stick together and to other surfaces, related to moisture content and particle size;
- Fibrous nature, as fibrous materials are a special case and must be comminuted in shredders or cutters e.g. based on the hammer mill design;
- Melting point: heat generated in a mill may be sufficient to cause melting, resulting in problems of increased toughness, cohesivity and adhesivity;
- Other special properties: thermally sensitive materials with a tendency to spontaneous combustion must be ground using an inert carrier medium, toxic or radioactive materials must be ground using a carrier medium in a closed circuit.

2.6 Carrier Medium

A carrier medium can be used to transport the material through the mill, though also to transmit forces to the particles. The medium also has an influence on friction, abrasion, crack formation and propagation, cohesivity and adhesivity, electrostatic charging and inflammability: it is therefore an important part of the milling process.

For particle size reduction, the carrier medium may be a gas or a liquid. Air is the most common used gas, except when an inert gas is mandatory when the material could ignite and/or be liable to dust explosion. In wet grinding, a liquid is used as carrier medium, with water as the most common, but oils can also be used.

2.7 Mode of Operation

Mills operate in batch or continuous throughput or capacity, and the choice between these two modes depends on throughput, process and economics. A batch mill has a capacity that varies from a few grammes at laboratory scale, to a few tonnes at commercial scale. The throughput of a continuous mill varies from several hundred grammes per hour at laboratory scale to several thousand tonnes per hour at industrial scale.

2.8 Types of Milling Circuit

Milling circuits can be either open or closed. In an open circuit, the particulate material passes through the mill only once, making the residence time the only controllable variable: product size and distribution can be controlled over a limited range by varying the throughput. In a closed circuit, the material being discharged is classified, and the oversized particles are returned to the mill together with new feed material. Both particle size and distribution can be more flexibly controlled.

3 Physico-Chemical Production of Fine Particles

This section reviews several physico-chemical techniques for the production of ultrafine powders and some practical applications are given.

3.1 Thermal Decomposition

Thermal decomposition or thermolysis is a chemical reaction in which a chemical substance breaks up into at least two chemical substances when heated. Heat is usually required to break chemical bonds, which makes thermolysis an endothermic reaction. Thermal decomposition is used to produce very fine particles. The process can be controlled via the rate of heating: fast heating will result in many nuclei and a very fine powder, whereas slow heating will result in few nuclei only producing coarser powder, but well defined crystals. Examples of the application of thermolysis are the production of MgO from $Mg(OH)_2$ at 400 °C, $BaTiO_3$ from $BaTiO(C_2O_4)0.4H_2O$ at 900 °C, and γ-Fe_2O_3 from α-$FeOOH$ at 300 °C.

3.2 Deposition from a Liquid

Deposition of solids from liquids can happen by precipitation, hydrolysis, hydrothermal reaction, solvent evaporation or rapid expansion of supercritical solutions (RESS).

The first method to produce ultrafine powder is the process of liquid phase *precipitation*. An example of an ultrafine powder produced through this technique is γ-Fe_2O_3:

$FeSO_4$ and NaOH react to form $Fe(OH)_2$, which is oxidized to α-FeOOH and heated to γ-Fe_2O_3. [11]

The production of α-Al_2O_3 is an example of *sol-gel* hydrolysi. This ultrafine powder has a considerable potential for a wide range of applications including high strength materials, electronic ceramics and catalysts. Because of its wide and common use, there has been an increasing interest in the synthesis of this powder. Al_2O_3 is a complex oxide with several meta-stable phases, which eventually convert to the stable α-Al_2O_3 phase. Several conventional production techniques are available, each with certain disadvantages: [1]

- Mechanical milling requires extensive ball milling and easily introduces impurities;
- Vapor phase reaction from a gas phase precursor is a very energy intensive process, as it requires temperatures above 1200 °C;
- Precipitation is very complex and very time consuming (long washing and aging times);
- Hydrothermal production requires high temperature and pressure, which again consumes much energy.

Because of these disadvantages, the most common used technique for the production of α-Al_2O_3 is sol-gel hydrolysis, making use of metal alkoxides as raw material. High prices of these alkoxides however, together with long gelation periods, limit the application of this method. An alternative, efficient, low-cost aqueous sol-gel method was developed by Li et al. The starting solution consists of aluminium nitrate and citric acid dissolved in deionised water. This solution is continuously stirred for several hours and kept at a temperature of 60 °C until it turns into a yellowish sol. The stabilized sol is then rapidly heated to 80 °C and constantly stirred. As the sol turns into a transparent gel, viscosity and color change. The gel is heat treated at 200 °C for 2 h, which results in a fluffy, polymeric citrate precursor. Finally, this precursor can be ground to a fine powder, the ultrafine α-Al_2O_3.

Hydrothermal techniques involve the dissolution of crystals of low solubility in water, by heating the solution above 100 °C, where after recrystallization occurs.

For *solvent evaporation*, two types of processes are used. When a solution of a high soluble metal salt is sprayed directly into a furnace, the water will evaporate and thermal decomposition will occur. This process is called *spray drying* and is illustrated in Fig. 12. Liquid droplets can also be sprayed into a very cold liquid like hexane, dry ice, or liquid N_2, a process called *freeze drying*. The droplets are frozen, the pressure is reduced and heat is supplied to remove water by sublimation.

Evaporation processes are dealt with in e.g. [12].

A special type of spray drying is the process of *spray pyrolysis*, illustrated in Fig. 13. Solutions containing the precursors of particles are sprayed, so tiny droplets are formed. At the surface of the liquid droplets, solvents are vaporized. Precursors are precipitated, followed by continuous decomposition and agglomeration in the reactor, until the desired products are obtained [11].

Rapid expansion of supercritical solutions, the so-called RESS process, starts from a homogeneous supercritical solution of solutes in supercritical CO_2. By reducing the density of this solution through expansion to atmospheric conditions, nucleation is rapidly induced. This nucleation results in a rapid decrease in solvent strength, causing high

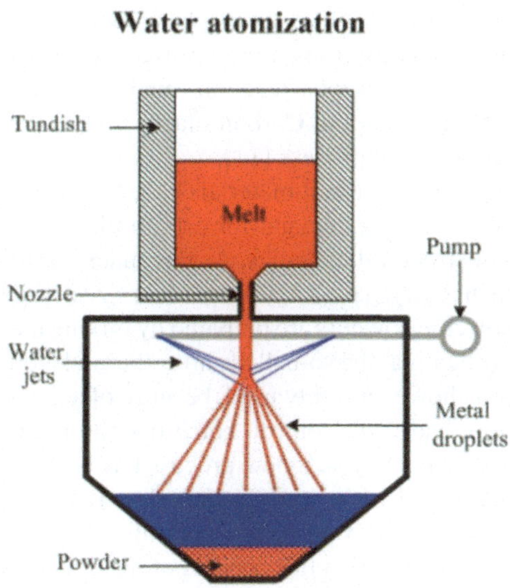

Fig. 12. Spray drying in powder metallurgy [13]

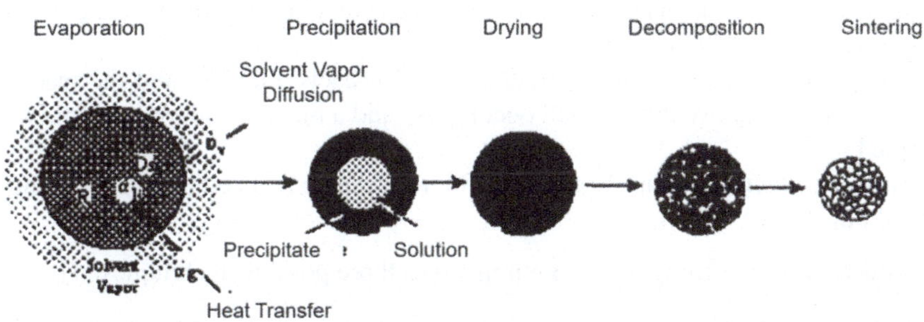

Fig. 13. Principle of spray pyrolysis [11]

supersaturation. This technique induces very high homogeneous nucleation rates. The nucleation and growth time are very short (10^{-5}–10^{-6} s), and create very small particles. RESS hereby offers a method for controlled homogeneous nucleation.

3.3 Vapour Phase Methods

The production of fine particles using vapour phase methods involves aerosol processes: particles are formed and grow in supersaturated vapours. These vapours can be produced either chemically (by reaction) or physically. Chemical vapour deposition is a gas phase reaction, which is suitable for preparing ultrafine particles with controlled shapes. This vapour phase method is used e.g. for the production of TiC from $TiCl_4$ reacting with CH_4. In physical vapour deposition, vapours are cooled rapidly, resulting in vaporization and aggregation of particles. The vapours can be formed either at high temperatures, high vacuum or after chemical reaction [1].

Production of ultrafine particles by vapour phase methods is essentially a process in which molecules and atoms in gases are converted into particles. This conversion passes through several stages: gaseous reaction, nucleation, particle growth, surface reaction, coagulation and aggregation. Carbon black, SiO_2 and TiO_2 are examples of the commercial application of the process [11].

High temperature and wear resistant materials have become increasingly important, and is illustrated by the increasing demand for certain ultrafine powders like BC_4 and SiC for the production of structural ceramics. As raw materials of novel products, these powders need to be high in purity, have a narrow particle size distribution and a very small particle size. Traditionally, the powders are prepared by mixing the corresponding oxides with carbon, followed by heating: the so-called carbo-thermal reduction process. In mass production, this process is however rarely used, because of some serious disadvantages: poor product distribution, bad form control, batch operation, no automation possible, etc. To overcome these disadvantages, a new process has been successfully developed. First, $SiCl_4$ is uniformly mixed with either heavy oil or B_2O_3 and C, and this mixture is immediately fed into the reactor. Within tens of milliseconds, the desired temperature of 1700–1900 °C is reached through a fast process of mass and heat transfer. Properties of BC_4 particles produced by this process meet the requirements [11].

Production of fine metal powders involves three possible techniques [1]:

- Evaporation and condensation: by cooling a zinc vapour, the vapour will condensate and zinc powder will be formed;
- Thermal decomposition: when a metal (Fe, Ni) reacts with CO and the reactant is heated, vapours above 200 °C will decompose and a metal powder (2–3 μm) will be formed;
- Hydrogen reduction: a vapour of metal chloride can be reduced by H_2, resulting in a metal powder (ca. 25 nm).

Production of fine compounds again involves three possible techniques [1]:

- Nitrogen plasma: by putting an arc of melting TiN in nitrogen plasma, a powder of TiN (20–300 nm) can be formed;
- Electric arc: when a high voltage discharge is applied to a metal held between carbon electrodes, a powder of ca. 50 nm is formed (TiC, TiS, etc.);
- Thermal hydrolysis: a vapour of metal chloride can be oxidized by O_2, resulting in a powder of ca. 16 nm (SiO_2, TiO_2, AlO_3, etc.).

Other techniques to produce fine particles using vapour phase methods are [1]:

- Hybrid plasma: subjection of a mixture of metal vapour and carrier gas to a high frequency plasma arc jet (Si_3N_4 at ca. 30 nm);
- High temperature reaction: reaction of a vapour of metal chloride with NH_3 at 1200 °C (WC, Mo_2C, etc. at 50–300 nm);
- High temperature pyrolysis: thermal destruction of hydrocarbons at 1000 °C (carbon black at 23 nm).

Illustrations are given in [14–17].

3.4 Solid State Reactions

Solid state reactions for the production of fine particles do not involve special processing: the fine particles are formed during the reaction. Some examples of solid state reactions are the production of (i) cement clinker from silica and lime, (ii) carbon black from the pyrolysis of carbohydrates, (iii) manganese ferrite for magnetic recording heads, and (iv) barium titanate for capacitors.

In the production process of *cement*, the raw materials are fed into a rotary kiln where they are heated to 1400–1500 °C. The raw feed enters the kiln at the cool end, gradually passes down to the hot end, is discharged from the kiln and cools down. In the temperature range of 1300–1450 °C, alite is formed by the spontaneous reaction of lime and silica. Fast cooling of this alite results in cement powder.

Pyrolysis of wooden biomass at approximately 380 °C results in a 95–99% conversion of the feed into fine carbon particles: this is the so-called *chartherm* process.

Driven towards smaller size, higher *capacitance* and lower cost, the manufacturing techniques of capacitors are undergoing a revolution, with the development of multilayer ceramic capacitors MLCCs. Production of MLCCs generally uses expensive metal powders, which can however be replaced by nickel powders. This ultrafine powder is also used in many other applications: conductive pastes, battery materials, catalysis materials, magnetic materials, etc. Wet chemical processes are considered to be the most suitable methods to obtain nickel powders, suitable for MLCCs: the powder is obtained by the hydrazine reduction of nickel sulfate in an ethanol-water solvent. Nickel sulphate is dissolved in the solvent and heated. Heated hydrazine is added to the solution, which reduces the nickel sulphate to nickel powder. This powder is separated from the solution by centrifugal separation, and washed with distilled water followed by ethanol and acetone. Finally, the powder is dried in a vacuum dry oven at 40 °C, so the gray-black nickel powder is obtained [18].

4 Conclusion

Fine and ultrafine particles are of growing importance for numerous industrial processes.

Ultrafine powders have widespread applications, due to their excellent properties: very high chemical activity due to their high specific surface up area), low melting and sintering point, high specific heat.

The different techniques were reviewed in this paper. Although mechanical production remains the main techniques, several other techniques are available and are reviewed. The energy consumption and obtained quality are often the selection criteria.

References

1. Brems, A., Baeyens, J., Dewil, R.: De productie van fijne en ultrafijne poeders. Kluwer Editorial (2009)
2. Lechanteur, A., Gresse, E., Orozco, L., Plougonven, E., Léonard, A., Vandewalle, N., Lumay, G., Evrard, B.: Inhalation powder development without carrier: how to engineer ultra-flying microparticles? Eur. J. Pharm. Biopharm. **84**, 30–44 (2023)

3. Tutor Vista.: www.tutorvista.com (2023)
4. Griffith, A.A.: Phil. Trans. R. Soc. **221**, 163 (1921)
5. Perry, R.H., Green, D.W.: Perry's Chemical Engineers' Handbook, 7th ed. McGraw-Hill International Editions (2007)
6. Rhodes, M.J.: Introduction to Particle Technology. Wiley (2008)
7. Northstone Materials.: www.northstone.com (2023)
8. British Geological Survey.: www.bgs.ac.uk (2023)
9. TechnoMine.: http://technology.infomine.com (2023)
10. Zenith Impact Crushers.: www.impact-crushers.com (2023)
11. Hu, L., Chen, M.: Preparation of ultrafine powder: the frontiers of chemical engineering. Chem. Ind. Eng. Progress **43**, 212–219 (1996)
12. Vladisavljević, G.T.: Preparation of microparticles and nanoparticles using membrane-assisted dispersion, micromixing, and evaporation processes. Particuology **84**, 30–44 (2024)
13. Substances and Technology.: www.substech.com (2023)
14. Eessaa, A.K., Elkady, O.A., El-Shamy, A.M.: Powder metallurgy as a perfect technique for preparation of Cu–TiO_2 composite by identifying their microstructure and optical properties. Sci. Rep. **13** (2023)
15. Hamidi, A., Shakibania, S., Mahmoudi, A., Rashchi, F., Vahidi, E.: Valorization of fly ash by nickel ferrite and vanadium oxide recovery through pyro-hydrometallurgical processes: technical and environmental assessment. J. Environ. Manag. **344**, 118442 (2023)
16. Han, W., Li, Y., Wang, Y., Xiaoming, N.I., Wang, L., Zhou, Y.: Characterization of coal fines and their production controlling factors: a case study from Southern Qinshui Basin, China. Nat. Resour. Res. **32**, 1777–1794 (2023)
17. Wang, Z., Cheng, Y., Li, K., Wang, T., Hu, M., Zheng, C., Li, X., Huang, Z., Hu, X., Li, Y.: Synthesis and formation mechanism of HfB2 ultrafine powders with low oxygen via flocculating settling assisted process and carbo/borothermal reduction. J. Mater. Sci. Technol. **164**, 229–239 (2023)
18. Chen, R.Y., Zhou, K.G.: Preparation of ultrafine nickel powder by wet chemical process. Trans. Nonferrous Metals Soc. China **16**, 1223–1227 (2006)

Energy Transformation and Sustainable Development

Vulnerable Groups and Social Marginalization Risks in the Digital Energy Transition: A Review

Philipp Thunshirn(✉) and Ines Mlinaric

University of Applied Sciences Technikum Wien, Giefinggasse 6, 1210 Vienna, Austria
philipp.thunshirn@technikum-wien.at

Abstract. Digitized renewable energy systems promise to promote accessibility and participation in energy markets by end users. While some use cases unlock new potentials for certain groups in society, they are not easily available for all. Unequal access to digital infrastructures and chances for participation, as well as differing digital and energy literacy, could inhibit vulnerable populations from participation in several use cases. This review lists use cases in smart energy systems from a user perspective, and pinpoints social strata that are at particularly high risk of being marginalized. Based on respective preconditions for participation, we deduce hypotheses for social marginalization risks. Sociodemographic factors such as income, level of education and age are critical for participation. In addition, digital literacy depends on socioeconomic status: A higher socioeconomic position leads to better digital literacy. End users' environmental awareness and trust are also decisive for participation in the digital renewable energy transition. Hence, end users from certain social milieus with less environmental awareness and stronger institutional mistrust, who are less integrated into the local community run higher risks of being further left behind. Looking through the lens of vulnerable end users, we find that the narrative of social inclusion is confronted with the conceivable social exclusion or marginalization. Consequently, energy policies should generally be scrutinized for more social justice and designed accordingly.

Keywords: digital energy · social exclusion · use cases

1 Introduction

The energy sector has been an early adopter of digital technologies, as they promise to solve the most pressing challenges of the renewable energy transition, such as integrating renewable energy sources, increasing energy efficiency, and strengthening energy supply security [1]. Intermittent and spatially distributed renewable generation capacities require greater flexibility in the energy system and increase its control and regulation requirements. According to the vision, digital ICTs and smart meters will enable precise measurements, control, and real-time trade of energy flows, which is considered key to adjusting household load profiles to fluctuating generation capacities [2, 3]. Digital innovations are reshaping the highly institutionalized, top-down organized energy system, and animating passive consumers to seize the option of becoming active participants in energy markets [4, 5].

J. Baeyens et al. (Eds.): ESRE 2023, pp. 183–200, 2025.
https://doi.org/10.1007/978-981-96-4345-5_14

However, the disruptive dynamics of digitization also increase complexity and effects decentralization in the supply chain, leading to role changes in the sectoral network of actors. Although ICTs and smart meters pledge to promote the accessibility, efficiency, and sustainability of renewable energy systems, they also raise concerns for end users such as affordability, operability, and privacy [6, 7]. While some use cases unlock new prospects for certain societal groups, the same might cause unintended negative side-effects for others that are left behind (digital divide) [8–12]. Unequal access to ICTs and devices plus uneven societal digital literacy required to use ICT may reinforce social inequality [10]. Consequently, vulnerable population groups could face higher digital exclusion and marginalization risks. Some use cases may even promote the social marginalization of end users based on their sociodemographic status. Yet, these concerns remain disregarded in current energy research.

This review article provides an inventory of use cases of digitization in the energy sector from an end user perspective. Driven by the approach that digital energy systems should develop under social (and ecological) premises [13], we ask who could be socially excluded or marginalized from these use cases and how. We analyze articles regarding end users' characteristics and use case acceptance to specify the personality traits of individuals that are prone to be left behind. Finally, we hypothesize social exclusion and marginalization risks of defined social strata. The aim of this review is thus not an empirical investigation of the reasons for social exclusion, but a literature basis for further research that facilitates addressing social equity.

The next section describes our review methodology. Subsequently, we describe digital use cases according to their functionalization. At the end of each subchapter, we recognize vulnerable groups. Finally, we hypothesize social exclusion and marginalization risks. Use cases from industry 4.0 such as additive manufacturing, 3D printing, and virtual reality, and the connectivity of the public and private transport sectors are beyond the scope of this study.

2 Methodology

Figure 1 gives a schematic overview of our review methodology. Our synthesis covers a literature body of 97 studies that are selected from a total of 527 articles from peer-reviewed journals available in the online databases of ScienceDirect, Taylor&Francis, and IEEE. The following keywords, as well as a combination of these, were deemed relevant to discover use cases of digitized energy systems:

- "digital" "smart" + "energy system" + "digital energy data"
- "smart energy systems" + "smart energy" + "digital power systems"
- "Digital divide energy" + "smart meter acceptance" + "smart home energy"

At first, we focused on finding use cases. Because the full scope became apparent after an initial search of the literature, various scientific review articles were examined, which provide a compilation of the comprehensive topic. Based on the findings in the review articles, sector-related use cases could be analyzed by employing a snowball system. After recognizing the most relevant use cases, we aggregated them according to their functionalization, because we consider this categorization as most practicable when investigating social equity from the perspective of vulnerable end users. Thereafter, we searched for social acceptance of the respective use cases to assemble acceptance factors.

- "digital energy monitoring" + "acceptance" + "exclusion"
- "energy cloud" + "acceptance" + "exclusion"
- "blockchain energy" + "acceptance" + "exclusion"
- "IoT energy" + "acceptance" + "exclusion"
- "digital demand side management/demand response" + "acceptance" + "exclusion"
- "smart home" + "acceptance" + "exclusion"

We reviewed sociodemographic and personality-related factors that allow us to identify vulnerable groups of digitization in energy systems. Based on these vulnerable groups, we hypothesized social exclusion and marginalization risks and deliberated existing research gaps for future studies.

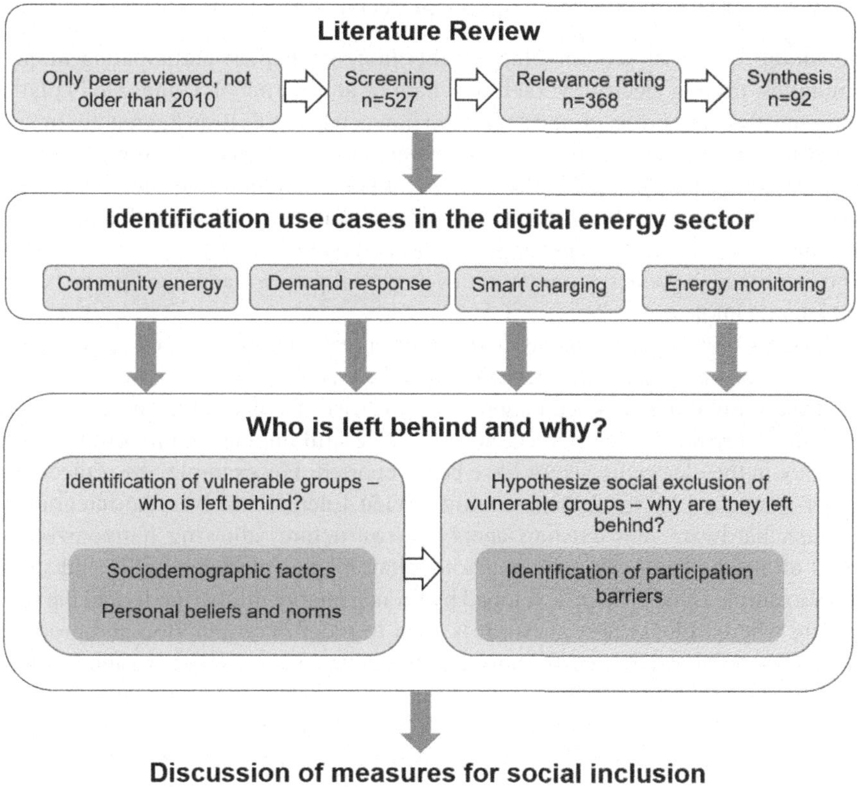

Fig. 1. Schematic overview of the methodical approach.

3 Use Cases Inventory and Vulnerable Groups

The next subchapters provide an overview of digital use cases in the energy sector. Comprehending that these use cases can also be categorized along other criteria (e.g. along the supply chain), we ask which benefit the respective use case can bring. Thus, the functionalization of the use cases is our focus when gauging their social equity impact. These categories are not intended to represent a strict separation because some use cases are a mix of several, overlap, and can also be assigned to others. Moreover, decentralized production and consumption of energy blur the boundaries of conventional top-down energy supply. Such a demarcation can, however, support evaluation and comparison. After the description of the use case, each chapter compiles studies of influencing factors that contribute to or suppress participation in the respective use case. Hence, vulnerable groups can be determined to infer social exclusion and marginalization risks hypotheses.

3.1 Use Case #1: A Person Participates in Peer-to-Peer (P2P) Electricity Trading

Three similar use cases are combined in this first subchapter: participating in energy communities, trading electricity via blockchain or an Internet of Things (IoT) platform in community energy projects, and supplying power capacities to an aggregator in virtual power plants. Prosumers distribute self-generated energy in their local grid, aiming to reduce CO_2 emissions through collective investments in renewable energy generation, thus generating economic benefits for end users (prosumers), and strengthening the local community via sharing concepts and awareness raising [14–16]. These use cases can also be combined, e.g. energy communities or virtual power plants can also be based on blockchain technology.

An energy community is the association of at least two participants, to jointly produce, store, consume and sell renewable electricity across property boundaries. Based on the legal regulations in some nations (e.g. in Austria), renewable energy communities in Austria are not profit-oriented. Several successful implementations of blockchain technology in the electricity sector have been reported. For example, the project "Viertel Zwei" launched by the Viennese utility Wien Energie involves the integration of blockchain hardware into existing energy infrastructure, allowing harmonization of services of all applications - such as photovoltaic plants, electromobility and storage. Another example is the project developed by French energy utility Enedis and the startup Sunchain, wherea blockchain platform is used to record consumption and production data, which is collected by smart meters and distribute energy among prosumers through smart contracts [17]. A project by Dutch transmission system operator Tennet and battery producer sonnen is using blockchain technology to stabilize the power grid. The project enables electric car owners to sell excess electricity generated by home photovoltaic systems to the power grid to balance the grid's energy needs [17, 18]. A virtual power plant is a collection of multiple distributed power generation, storage, and controllable or flexible loads that form a virtual unit. Decentralized technologies such as photovoltaic, electricity storage, and electric vehicles are organized from the power grid [2, 3]. Typical examples of virtual power plants are the "PowerMatcher" project in Hoogkerk, Netherlands [19], a project by the utility Fortum in collaboration with the company Ericsson in Finland [20], and the "sonnenCommunity" [2, 3].

Vulnerable groups. The typical participant of P2P trading is male, well-educated and has a higher income, lives in a rural area, has strong pro-environmental behaviors as well as a predisposition for active citizenship [21–23]. Other studies found, that investing in community energy projects is relatively stable across gender, age, and employment [24–26]. However, Hanke et al. [27] found that member diversity and access to decision making in energy communities in Germany are dominated by male participants [27]. Fraune [28] also identified a significantly lower share of women in leadership positions in energy communities and a lower female ownership rate and investment volume [28]. Therefore, participating in energy communities may reinforce gender disparities (which can, however, also be a result of existing general gender disparities). There is a positive effect of home and energy generation ownership (e.g. rooftop PV) on readiness to invest in P2P electricity trading [25, 29]. However, Conradie (2021) found insufficient evidence that income is significantly related to willingness to invest [25].

Environmental awareness is also the main driver to participate in P2P trading [21, 22, 25, 29–32]. Moreover, a positive attitude toward innovations in general, familiarity with energy topics [33], and contribution to societal energy autarky [34] foster participation. Further studies found by factor and multi-variate regression analysis that environmental concern, energy independence, trust, higher education, and awareness about local energy initiatives are fundamental drivers of the citizens' willingness to participate in energy communities [24, 29]. End users who have a positive attitude towards the renewable energy transition are generally more willing to participate in energy communities. Certain social classes with certain values and attitudes are eager to participate, potentially leaving others behind.

3.2 Use Case #2: A Person Participates in Demand Response Activities

Due to the increasing share of volatile, renewable energy sources in the power grid, the equilibrium between energy demand and generation could be seriously disturbed in the future. In this context, households can provide demand-side flexibility by changing their demand in response to a given (price) signal. This refers to the ability of consumers to change their consumption level and the timing of demand of energy-intensive and frequently used household appliances (washing machines, dishwashers, tumble dryers, heat pumps, electric boilers, etc.) [35, 36]. Thus, grid stability improves and peak loads are reduced (which are often covered by expensive and CO_2-intensive gas turbines, also at high shares of fluctuating renewable energy generation. End users are encouraged to postpone household practices to take advantage of variable tariffs, thus gaining monetary profits and decreasing CO_2 emissions through participation [37]. Customers authorize the utility to schedule, reduce, or shut off household electricity use. For a more flexible heat and cooling supply, digital controllers can communicate via a cloud to raise heat supply flexibility [38]. Targets, such as temperature amplitudes, are set by making decisions based on internal system information and external measured influencing factors such as energy prices, temperatures, etc. By monitoring temperatures, processing real-time data and controlling thermostats, peak thermal energy consumption can be reduced without significantly affecting thermal comfort in buildings [39]. In addition, connected buildings can also serve as mutual thermal energy storage systems [40].

Vulnerable Groups. Many households are still organized in a heteronormative manner: Men take up repair work and financial responsibilities whilst women predominantly command domestic routines (washing, cleaning, cooking) and childcare [41–43]. Thus, female end users with relatively stable daily routines are more affected by shifting household loads and avoiding peak times under variable pricing [44]. This may result in higher energy prices for women, particularly for single parents. Moreover, women are less likely familiar with technology and infrastructure [41, 44–46]. Hence, women could face higher barriers to adopt demand response activities, resulting in higher energy costs for them. Typically, younger generations are more informed about the meaning of smart grids and demand response. Thus, they are more willing to defer their household loads than older people. Further, higher living space and a higher number of household members increase inflexibility, resulting in a lower willingness to shift peak loads [47]. End users from low-income households may also be excluded from profitable load shifting and participation due to affordability and lack of access to necessary equipment [47]. Additionally, residents of well-insulated households capable to store heat longer are more likely to participate in thermal load shifting. Furthermore, lacking affordability of recent (and more expensive) smartphones or user interfaces with appropriate technical capabilities or inadequate Internet capacity excludes end users from participating [41, 44–46]. Low-income households are also vulnerable to poverty if they have to bear the cost of the required metering and control equipment without gaining notable profits [48]. Additionally, ownership of renewable sources of energy (e.g. PV Panel on the roof) provokes changes in demand response behavior [49, 50]. In turn, this signifies that those who cannot afford PV ownership are more prone to be left behind.

Similar to P2P trading, economic motives and environmental awareness, such as climate change risk perceptions, are the most common motivations for participation in demand response activities [47, 51–53]. In addition, emphasizing the promotion of energy independence and societal benefits can help to increase the willingness to participate [54]. Such normative aspects often have a greater impact on user participation than economic benefits [55, 56], especially among residents of apartments compared to those in single-family homes [57]. Some end users mistrust utilities, and this can be exacerbated by unfamiliarity with demand response and how it can create cost savings. Concerns around privacy and autonomy as well as opaque and less predictable dynamic pricing or automation schedules may erode trust over time associated with direct load control [51, 58], particularly among those already concerned about affordability [55]. Perceived lack of control over utility-controlled load-shifting remarkably reduces participation as a result of mistrust and the impression of privacy intrusion [59]. Stenner et al. [60] conducted a study where they sent trust-restoring messages to households upfront demand response activities. Willingness was increased compared to a control group, indicating that trust is key when participating in demand management services [60]. Automation or direct load control (loads are automatically controlled by the utility) may lessen the complexity and effort associated with responding, but participation may be discouraged by perceived risks and control by users [51].

3.3 Use Case #3: A Person Participates in Smart Charging and Charges/Discharges His/Her Electric Vehicle According to Renewable Energy Production

Similar to demand response, smart charging of electric vehicles intends to charge at times of high renewable energy production and low electricity demand and, vice versa, to discharge during high electricity demand and low renewable energy production. Here, CO_2 emissions are reduced because peak loads are often covered by gas power plants fueled by natural gas. Demand-side balancing can also be achieved through sector coupling of electricity and mobility, where electricity surpluses can charge electric vehicles while using batteries as storage media. Several successful implementations of smart charging have been reported, such as the Danish EDISON project or the AutoBidder platform [61].

Vulnerable Groups. Owners of electric vehicles and smart charging enthusiasts usually own detached homes with adequate recharge access and have higher levels of income and education [62–67]. Individuals with lower education levels living in rural areas typically have to cover larger distances to work and may have less battery capacities at a given range [67]. Similar to the other use cases, environmental awareness, perceived advantages to society, and technology-savviness are crucial factors for participation [62, 63, 68]. Contributing to grid stability and the integration of renewable energy sources is frequently an even more important factor than receiving economic value [69, 70]. However, environmental benefits alone will not succeed in convincing drivers to use smart charging [71, 72]. Above all, financial and socio-ecological aspects are the central motivating factors for drivers to use such smart charging systems [73–75], while range anxiety is a key factor in bidirectional charging (discharging electricity to the grid for ancillary services) [32]. Because of perceived personal control and lower perceived risk that a vehicle might not be fully charged at a certain time (e.g. middle of the night), end users prefer self-controlled charging over utility-managed charging. End users could attach substantial weight to risks of unlikely middle-of-the-night journeys (e.g. hospital emergency). A certain degree of automation of the end user to have ultimate control to override, therefore, fosters participation. Moreover, perceived additional daily effort is likewise unconducive [21].

3.4 Use Case #4: A Person Continuously Monitors His/Her Household Energy Consumption for Optimized and Conscious Energy Consumption

The purpose of energy monitoring of household appliances is to increase energy efficiency through optimized energy consumption. Digitalization enables real-time monitoring of energy consumption through connected digital sensors, often monitored in an IoT system [76]. Monitoring the energy data, analyzing it, and taking action based on it can enhance the energy efficiency of buildings. By installing smart plugs, which can measure the energy consumption of devices and automatically reduce their power if necessary, the overall power consumption of devices can be smart controlled, visualized, and reduced [77]. In addition, by measuring temperature, energy consumption, human presence via infrared sensors, and lighting, heating and cooling in buildings can be better adjusted and therefore used more efficiently, resulting in energy savings [76].

For data collection, energy consumption data is collected from every device in buildings and transmitted to a central server for further processing and analysis [78]. Individual smartphone-assisted WiFi tracking and predictive heating control algorithms can also be used to improve energy management [79, 80]. Through real-time energy consumption data collection via smart meters, a building's demand for heating and cooling, depending on building type, weather forecasts, and associated heating or cooling demand is estimated and controlled via a central cloud [81, 82]. Additionally, monitoring household's energy consumption can show residents how much energy is consumed at what time, resulting in behavioral changes and energy savings [41].

Vulnerable Groups. Comparable to the other use cases described above, enthusiasts of household energy monitoring are most likely men, with an above-average income, below 55 years of age, have a higher level of education and are technology-savvy [46, 83]. Additionally, low-income households facing increased complexity of smart technologies are often less engaged with energy issues and technology [45]. In recent studies about smart homes, men and technology enthusiasts were more likely to try the devices and very few people over 55 years of age were interested in or able to use them [41, 44–46]. The study by Brown and Markusson [84] showed that older adults were generally more aware of their energy use, also without energy monitoring. They already practiced energy-saving behaviors learnt from upbringing, resulting in negligible positive benefits and low engagement with the device. Moreover, lacking technical skills and confidence, and the risk of losing comfort and convenience, the device of energy monitoring triggered negative emotions amongst some older adults. [84].

Similar to the other use cases, environmental concerns, such as climate change risk perceptions, trust and privacy issues, and economic motives are the most mentioned in available literature [56, 83, 85, 86]. Higher levels of environmental awareness, which is often connected to higher levels of education and income [10], and economic motives are also the most important determents for participating in energy monitoring [87]. Perceived climate change risks and awareness of energy use are even more important than gaining monetary profits from energy monitoring. Contrary, economic motives are more important for citizens of lower socioeconomic status [83]. Also here, hedonic factors are relevant [86, 88].

4 Use Cases and Societal Equity

In the following, we hypothesize the risks of being socially excluded or marginalized. We understand exclusion as the lack of financial and infrastructural capabilities for participation. Marginalization denotes that participation is principally possible, but discourages end users from participation because of personal beliefs, awareness, energy and digital literacy and cognitive capacities. We argue, that both, exclusion and marginalization of vulnerable groups, contribute to social inequality.

Based on our literature findings, Fig. 2 demonstrates exclusion and marginalization. Dotted boxes represent (mostly financial) preconditions. The thickness of the colored arrows represents the importance of the emotional and cognitive factors of participation in the use case. The figure shows what are necessary and favorable factors for participation.

4.1 Social Exclusion Risks

We find that the social exclusion risks are mostly income-related. Lower-income individuals simply cannot afford the infrastructural investments to participate or take advantage of the use cases. We hypothesize the following social exclusion risks:

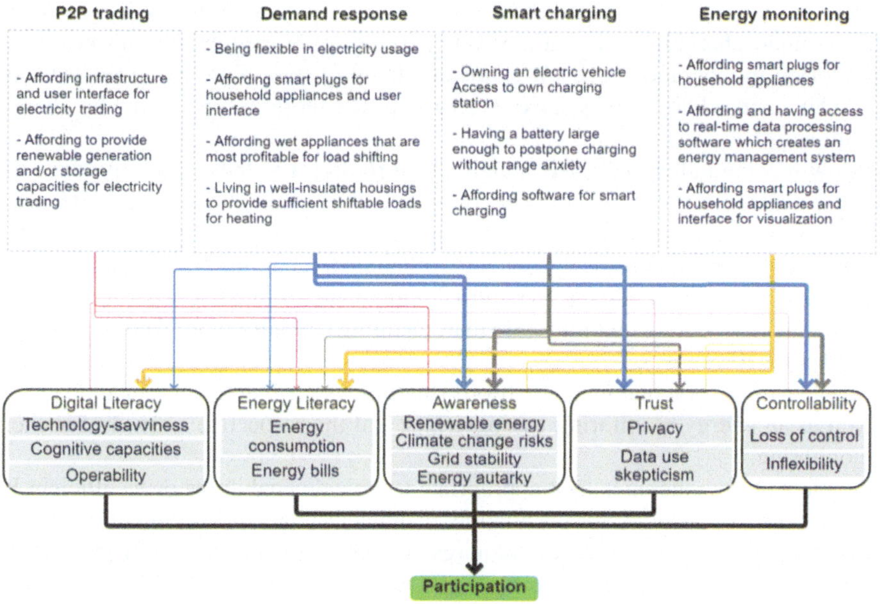

Fig. 2. Influencing factors to participation. Thicker arrows represent a stronger influence. The figure is based on literature results

- As most P2P trading use cases demand investments in local self-electricity generation, people with lower incomes are unable to finance rooftop PV and storage for P2P trading, hindering them to be active participants in community energy projects.
- Moreover, they are not capable to spend money on user interfaces that are often required to participate.
- Families with children and households with a higher number of household members are inflexible in shifting their chore-doing to another period. If inflexible residents are enrolled in variable tariff programs and cannot shift their loads from peak to off-peak times, they must pay an electricity higher price and cannot take advantage of the use case, especially female single parents.
- Further, end users from low-income households who cannot afford relevant electrical appliances (washing machine, dishwasher, tumble dryer) and the necessitated smart plugs and monitoring device plus software and installation are excluded from participation.
- Because load shifting of thermal energy for heating (e.g. shifting the load of the heat pump) requires good building insulation, end users who are living in poorly

insulated houses are excluded from gaining significant profits from thermal energy load shifting, meaning that they cannot reduce peak heat demand as effectively as people living in passive houses.

- People with low incomes cannot afford an electric vehicle at all to participate in smart charging/discharging.
- If, however, electric vehicles become eventually affordable to most people, they still face substantial barriers to smart charging. Since smart charging needs a costly personal smart charging station at a private parking space, people with lower incomes who lack money and space for a private charging station are excluded from participation. In addition, they have to charge at the more expensive public charging stations.
- If low-income individuals also cannot afford rooftop PV, they cannot produce their own cheaper electricity for smart charging, resulting again in higher charging prices.
- Furthermore, low-income individuals buy used electric vehicles with low battery capacity first and, therefore, are unable to allow smart charging without range anxiety. As less-educated individuals often must commute far distances to work in rural areas, they can provide less flexibility for smart charging (because more battery capacity is needed for travelling).
- People with low incomes cannot afford smart plugs for their electrical appliances as well as an energy monitoring user interface and the respective software for energy monitoring.
- People with low incomes lack appropriate technical capabilities or insufficient ICT connection and capacity for installation, maintenance, and utilization of the devices. For example, energy monitoring requires enough internet capacity for precise measurements, control, and data processing of real-time energy flows. Moreover, digital smartphone apps to allow controlling the energy data exchange also need considerable computing power, mobile data, and storage space on their cell phones to enable the utilization. People with lower incomes may not be able to provide the necessary technical infrastructure to participate in the use case.

4.2 Social Marginalization Risks

Income determines the possibility if people can ultimately afford participation or not. However, other aspects are important for acceptance, that should be used to formulate hypotheses for being left behind, such as the following:

- People having low levels of environmental concerns such as low climate change risk perceptions, are not inspired to participate in any of the use cases. Financial rewards for participation might be low, and the effort to engage is not in their interest. Because lower environmental concerns are associated with lower education levels, citizens of lower socioeconomic status could be left behind.
- Because trust is key to engaging and enjoying participation, people of distrust towards devices, energy utilities and grid operators do not join. Skepticism about data use and subjective loss of privacy might be major inhibitors to participation. Because distrust in public utilities (and authorities) is often associated with lower socioeconomic status, they could be marginalized.

- People, predominantly women, being less technology-savvy are not motivated to engage with the use case. They simply do not want to spend their leisure learning the function of the technology, as economic benefits might not outweigh the effort to install and use the system, and environmental protection is not a relevant motivator.
- Compared to younger people, older adults have not grown up with modern, digital technologies in their daily lives. Older adults might be not competent to operate digital technologies and lack the technical skills and cognitive abilities to participate in the use cases.
- The blockchain technology is too abstract to the average end user. Particularly less-educated and older adults lack familiarity with the term and functioning of the blockchain technology, the benefits it brings, and how energy can be traded via the blockchain. Engaging requires significant knowledge and learning. Those who cannot understand the technology could be less motivated to participate.
- Particularly older and less-educated citizens may not understand the societal benefits of the use cases. As younger, well-educated citizens are more aware of the ongoing renewable energy transition, climate change risks, and digital energy innovations, older and less-educated adults might not grasp the vision behind it. Contributing to grid stability and societal energy autarky, enhancing the integration of renewable energy sources, and climate change mitigation requires interest of the renewable energy transition and understanding the function of energy systems and its impacts on climate change. This may be too abstract for them. As economic benefits may not be the first motivator for participation, their enthusiasm to engage in the use cases could be too low.
- People with lower education have usually a lower energy literacy. They are not aware of their energy bills and energy consumption, lack knowledge of energy-intensive household practices, and how the energy supply chain (production, distribution, and consumption) works in general, which is an important factor when engaging in demand response activities, P2P trading or energy-efficient behavior.
- Older adults already practice perceived energy-saving behaviors in their households because they have learned it already from upbringing. Installing an energy monitoring and energy efficiency system would not bring a noteworthy perceived benefit to them. However, their measures to save energy consumption may be ineffective, hence, preventing them to exploit the full potential of the digital energy monitoring feedback system.
- People perceive the loss of control of their household appliances. Again, economic benefits may not outweigh their desire to control their chore-doing or heating at all times of the day and environmental protection is not relevant enough. Also, the impression of being controlled by somebody else (or by artificial intelligence or algorithms) and restricted in daily routines during utility-controlled demand side management, causes discomfort.
- Because hedonic motivations are also relevant for participation, individuals with low hedonic motivations are less likely to participate. Also, people who are rather passive citizens that are not integrated well into the local community may perceive themselves as societal outsiders who do not want to engage in community (energy) projects.

- As participating in renewable energy communities (in Austria) is non-profit-oriented, individuals of lower socioeconomic status that are motivated by financial incentives, are discouraged to participate.

The results of Hidalgo et al. [10], who assessed the digital divide, are in broad agreement with our review findings. They found that socioeconomic factors such as income, education level and age determine the digital divide (which is defined as unequal access to and use of ICTs). Also, digital literacy depends on socioeconomic status: higher socioeconomic status yields superior digital literacy. Likewise, Chadwick et al. [89], who conducted a comprehensive review of influencing factors for digital energy adoption and rejection and grouped them by relevance, found similar results to this work. Females and persons with lower levels of education and income are more likely to reject participation in digital energy innovations. The average end user lacks knowledge and understanding of the function of the use cases, have lower levels of environmental concern, and is more motivated by financial rewards [89]. Trust and confidence are, moreover, crucial for end users to participate in the digital energy transition [56, 90, 91].

5 Discussion and Conclusion

Published studies evaluating digitization in the energy sector lack analyzing the effects on social equity and social justice. Although existing research addressed the acceptance and the factors why energy innovations are supported and adopted or not, exclusion and marginalization risks are rarely considered. Unintended negative side-effects when introducing digital innovations in the energy sector, that aim to engage people in the energy transition (such as the use cases described above), should be recognized at an early stage and included in the respective energy policy design.

Above all, there is a lack of reasons for people´s rejection. Our hypotheses on why people are socially excluded can therefore provide a basis for future research to falsify (or verify) them. Such an investigation would give further insight into why people do not participate and provide a basis for energy policies to foster social justice in the energy transition. Acceptance should not be the sole criterion for whether the innovation leads to more social equality. Only because the innovation is accepted does not mean that all levels of society can benefit or participate. Some vulnerable populations may want to participate but are unable to. The focus of future research should not be on elaborating on whether people want to participate, but on whether they believe they actually can participate, and if not, to explore non-participation and its reasons in more depth.

Finally, we consider the general existing social inequality in societies, which produces vulnerable groups, as a root cause of differing prerequisites for participation. The fact that low-income individuals are at severely higher risk of being digitally left abandoned from the use cases described above, comes from existing income inequalities in societies. We argue that addressing income inequality as a whole can be a measure to reduce social exclusion and marginalization risks through the digitization of energy systems. Generally, we believe that a dialectical viewpoint can provide valuable insights into whether use cases can contribute to achieving desirable goals, or may even reinforce existing disparities.

References

1. IEA.: Digitalization and Energy (2017)
2. Liu, C., Yang, R.J., Yu, X., Sun, C., Wong, P.S., Zhao, H.: Virtual power plants for a sustainable urban future. Sustain. Cities Soc. **65**, 102640 (2021). https://doi.org/10.1016/j.scs.2020.102640
3. Rouzbahani, H.M., Karimipour, H., Lei, L.: A review on virtual power plant for energy management. Sustain. Energy Technol. Assess. **47**, 101370 (2021). https://doi.org/10.1016/j.seta.2021.101370
4. Botelho, D.F., Dias, B.H., de Oliveira, L.W., Soares, T.A., Rezende, I., Sousa, T.: Innovative business models as drivers for prosumers integration—enablers and barriers. Renew. Sustain. Energy Rev. **144**, 111057 (2021). https://doi.org/10.1016/j.rser.2021.111057
5. Niesten, E., Alkemade, F.: How is value created and captured in smart grids? A review of the literature and an analysis of pilot projects. Renew. Sustain. Energy Rev. **53**, 629–638 (2016). https://doi.org/10.1016/j.rser.2015.08.069
6. Lange, K., Knieling, J.: EU smart city lighthouse projects between top-down strategies and local legitimation: The case of Hamburg. Urban Plann. **5**(1), 107–115 (2020). https://doi.org/10.17645/up.v5i1.2531
7. Xu, Q., Zhong, M.: The impact of income inequity on energy consumption: the moderating role of digitalization. J. Environ. Manag. **325**(Pt A), 116464 (2022). https://doi.org/10.1016/j.jenvman.2022.116464. PMID: 36242971
8. Wang, S., Cao, A., Wang, G., Xiao, Y.: The Impact of energy poverty on the digital divide: the mediating effect of depression and Internet perception. Technol. Soc. **68**, 101884 (2022). https://doi.org/10.1016/j.techsoc.2022.101884
9. Roberts, S.: Making 'No one left behind' meaningful in our future energy system [cited 2022 November 18] (2018). Available from: https://www.cse.org.uk/news/view/2281
10. Hidalgo, A., Gabaly, S., Morales-Alonso, G., Urueña, A.: The digital divide in light of sustainable development: an approach through advanced machine learning techniques. Technol. Forecast. Soc. Chang. **150**, 119754 (2020). https://doi.org/10.1016/j.techfore.2019.119754
11. Liao, S.-C., Chou, T.-C., Huang, C.-H.: Revisiting the development trajectory of the digital divide: a main path analysis approach. Technol. Forecast. Soc. Chang. **179**, 121607 (2022). https://doi.org/10.1016/j.techfore.2022.121607
12. Bielig, M., Kacperski, C., Kutzner, F., Klingert, S.: Evidence behind the narrative: critically reviewing the social impact of energy communities in Europe. Energy Res. Soc. Sci. **94**, 102859 (2022). https://doi.org/10.1016/j.erss.2022.102859
13. Gährs, S., Aretz, A., Rohde, F., Zimmermann, H.: Die Digitalisierung des Energiesys-tems muss unter nachhaltigen Prämissen erfolgen (2021)
14. Sousa, T., Soares, T., Pinson, P., Moret, F., Baroche, T., Sorin, E.: Peer-to-peer and community-based markets: a comprehensive review. Renew. Sustain. Energy Rev. **104**, 367–378 (2019). https://doi.org/10.1016/j.rser.2019.01.036
15. Tushar, W., Yuen, C., Saha, T.K., et al.: Peer-to-peer energy systems for connected communities: a review of recent advances and emerging challenges. Appl. Energy **282**, 116131 (2021). https://doi.org/10.1016/j.apenergy.2020.116131
16. Zhang, C., Wu, J., Long, C., Cheng, M.: Review of existing peer-to-peer energy trading projects. Energy Procedia **105**, 2563–2568 (2017). https://doi.org/10.1016/j.egypro.2017.03.737
17. Wang, Q., Li, R., Zhan, L.: Blockchain technology in the energy sector: from basic research to real world applications. Comput. Sci. Rev. **39**, 100362 (2021). https://doi.org/10.1016/j.cosrev.2021.100362
18. Aretz, A., Knoefel, J., Gährs, S.: Prosumer-Potenziale in NRW 2030 (2017)

19. Milchram, C., Künneke, R., Doorn, N., van de Kaa, G., Hillerbrand, R.: Designing for justice in electricity systems: a comparison of smart grid experiments in the Netherlands. Energy Policy **147**, 111720 (2020). https://doi.org/10.1016/j.enpol.2020.111720
20. Fortum.: Ericsson's data centre and Fortum to collaborate in demand response. Fortum. Available from: https://www.fortum.com/media/2018/01/ericssons-data-centre-and-fortum-collaborate-demand-response [cited 2021 February 15]
21. Ebers Broughel, A., Hampl, N.: Community financing of renewable energy projects in Austria and Switzerland: profiles of potential investors. Energy Policy **123**, 722–736 (2018). https://doi.org/10.1016/j.enpol.2018.08.054
22. Schall, D.L.: More than money? An empirical investigation of socio-psychological drivers of financial citizen participation in the German energy transition. Cogent Econ. Fin. **8**(1), 1777813 (2020). https://doi.org/10.1080/23322039.2020.1777813
23. Radtke, J., Ohlhorst, D.: Community energy in Germany—bowling alone in elite clubs? Util. Policy **72**, 101269 (2021). https://doi.org/10.1016/j.jup.2021.101269
24. Koirala, B.P., Araghi, Y., Kroesen, M., Ghorbani, A., Hakvoort, R.A., Herder, P.M.: Trust, awareness, and independence: insights from a socio-psychological factor analysis of citizen knowledge and participation in community energy systems. Energy Res. Soc. Sci. **38**, 33–40 (2018). https://doi.org/10.1016/j.erss.2018.01.009
25. Conradie, P.D., de Ruyck, O., Saldien, J., Ponnet, K.: Who wants to join a renewable energy community in Flanders? Applying an extended model of theory of planned behaviour to understand intent to participate. Energy Policy **151**, 112121 (2021). https://doi.org/10.1016/j.enpol.2020.112121
26. Proudlove, R., Finch, S., Thomas, S.: Factors influencing intention to invest in a community owned renewable energy initiative in Queensland, Australia. Energy Policy **140**, 111441 (2020). https://doi.org/10.1016/j.enpol.2020.111441
27. Hanke, F., Guyet, R., Feenstra, M.: Do renewable energy communities deliver energy justice? Exploring insights from 71 European cases. Energy Res. Soc. Sci. **80**, 102244 (2021). https://doi.org/10.1016/j.erss.2021.102244
28. Fraune, C.: Gender matters: women, renewable energy, and citizen participation in Germany. Energy Res. Soc. Sci. **7**, 55–65 (2015). https://doi.org/10.1016/j.erss.2015.02.005
29. Kalkbrenner, B.J., Roosen, J.: Citizens' willingness to participate in local renewable energy projects: the role of community and trust in Germany. Energy Res. Soc. Sci. **13**, 60–70 (2016). https://doi.org/10.1016/j.erss.2015.12.006
30. Dóci, G., Vasileiadou, E.: "Let's do it ourselves" Individual motivations for investing in renewables at community level. Renew. Sustain. Energy Rev. **49**, 41–50 (2015). https://doi.org/10.1016/j.rser.2015.04.051
31. Bauwens, T., Gotchev, B., Holstenkamp, L.: What drives the development of community energy in Europe? The case of wind power cooperatives. Energy Res. Soc. Sci. **13**, 136–147 (2016). https://doi.org/10.1016/j.erss.2015.12.016
32. Bauwens, T., Devine-Wright, P.: Positive energies? An empirical study of community energy participation and attitudes to renewable energy. Energy Policy **118**, 612–625 (2018). https://doi.org/10.1016/j.enpol.2018.03.062
33. Seidl, R., von Wirth, T., Krütli, P.: Social acceptance of distributed energy systems in Swiss, German, and Austrian energy transitions. Energy Res. Soc. Sci. **54**, 117–128 (2019). https://doi.org/10.1016/j.erss.2019.04.006
34. Ecker, F., Hahnel, U.J.J., Spada, H.: Promoting decentralized sustainable energy systems in different supply scenarios: the role of Autarky aspiration. Front. Energy Res. **5**, 14 (2017). https://doi.org/10.3389/fenrg.2017.00014
35. Shariatzadeh, F., Mandal, P., Srivastava, A.K.: Demand response for sustainable energy systems: a review, application and implementation strategy. Renew. Sustain. Energy Rev. **45**, 343–350 (2015). https://doi.org/10.1016/j.rser.2015.01.062

36. Tarish Haider, H., See, O.H., Elmenreich, W.: A review of residential demand response of smart grid. Renew. Sustain. Energy Rev. **59**, 166–178 (2016). https://doi.org/10.1016/j.rser.2016.01.016
37. Elma, O., Selamoğullari, U.S.: An overview of demand response applications under smart grid concept (2017)
38. Guelpa, E., Verda, V.: Demand response and other demand side management techniques for district heating: a review. Energy **219**, 119440 (2021). https://doi.org/10.1016/j.energy.2020.119440
39. Salo, S., Jokisalo, J., Syri, S., Kosonen, R.: Individual temperature control on demand response in a district heated office building in Finland. Energy **180**, 946–954 (2019). https://doi.org/10.1016/j.energy.2019.05.035
40. Kensby, J., Trüschel, A., Dalenbäck, J.-O.: Potential of residential buildings as thermal energy storage in district heating systems—results from a pilot test. Appl. Energy **137**, 773–781 (2015). https://doi.org/10.1016/j.apenergy.2014.07.026
41. Hargreaves, T., Nye, M., Burgess, J.: Making energy visible: a qualitative field study of how householders interact with feedback from smart energy monitors. Energy Policy **38**(10), 6111–6119 (2010). https://doi.org/10.1016/j.enpol.2010.05.068
42. Tjørring, L., Jensen, C.L., Hansen, L.G., Andersen, L.M.: Increasing the flexibility of electricity consumption in private households: does gender matter? Energy Policy **118**, 9–18 (2018). https://doi.org/10.1016/j.enpol.2018.03.006
43. Chambers, D.: Attuning smart home scripts to household and energy care. Build. Cities **3**(1), 663–676 (2022). https://doi.org/10.5334/bc.220
44. Nicholls, L., Strengers, Y.: Peak demand and the 'family peak' period in Australia: understanding practice (in)flexibility in households with children. Energy Res. Soc. Sci. **9**, 116–124 (2015). https://doi.org/10.1016/j.erss.2015.08.018
45. Nicholls, L., Strengers, Y., Herreo Tirado, S.: Smart home control: exploring the potential for off-the-shelf enabling technologies in energy vulnerable and other households. Beyond Behav. Res. Program. (2017)
46. Tirado Herrero, S., Nicholls, L., Strengers, Y.: Smart home technologies in everyday life: do they address key energy challenges in households? Curr. Opin. Environ. Sustain. **31**, 65–70 (2018). https://doi.org/10.1016/j.cosust.2017.12.001
47. Li, R., Dane, G., Finck, C., Zeiler, W.: Are building users prepared for energy flexible buildings?—a large-scale survey in the Netherlands. Appl. Energy **203**, 623–634 (2017). https://doi.org/10.1016/j.apenergy.2017.06.067
48. Kim, J.-H., Shcherbakova, A.: Common failures of demand response. Energy **36**(2), 873–880 (2011). https://doi.org/10.1016/j.energy.2010.12.027
49. McKenna, E., Thomson, M.: Demand response behaviour of domestic consumers with photovoltaic systems in the UK: an exploratory analysis of an internet discussion forum. Energy, Sustain. Soc. **4**(1) (2014). https://doi.org/10.1186/2192-0567-4-13
50. Roth, L., Lowitzsch, J., Yildiz, Ö., Hashani, A.: Does (co-)ownership in renewables matter for an electricity consumer's demand flexibility? Empirical evidence from Germany. Energy Res. Soc. Sci. **46**, 169–182 (2018). https://doi.org/10.1016/j.erss.2018.07.009
51. Parrish, B., Heptonstall, P., Gross, R., Sovacool, B.K.: A systematic review of motivations, enablers and barriers for consumer engagement with residential demand response. Energy Policy **138**, 111221 (2020). https://doi.org/10.1016/j.enpol.2019.111221
52. Broberg, T., Persson, L.: Is our everyday comfort for sale? Preferences for demand management on the electricity market. Energy Econ. **54**, 24–32 (2016). https://doi.org/10.1016/j.eneco.2015.11.005
53. Pfeiffer, C., Puchegger, M., Maier, C., Tomaschitz, I.V., Kremsner, T.P., Gnam, L.: A case study of socially-accepted potentials for the use of end user flexibility by home energy management systems. Sustainability **13**(1), 132 (2021). https://doi.org/10.3390/su13010132

54. Parag, Y.: Which factors influence large households' decision to join a time-of-use program? The interplay between demand flexibility, personal benefits and national benefits. Renew. Sustain. Energy Rev. **139**, 110594 (2021). https://doi.org/10.1016/j.rser.2020.110594
55. Spence, A., Demski, C., Butler, C., Parkhill, K., Pidgeon, N.: Public perceptions of demand-side management and a smarter energy future. Nat. Clim. Change **5**(6), 550–554 (2015). https://doi.org/10.1038/nclimate2610
56. Balta-Ozkan, N., Boteler, B., Amerighi, O.: European smart home market development: public views on technical and economic aspects across the United Kingdom, Germany and Italy. Energy Res. Soc. Sci. **3**, 65–77 (2014). https://doi.org/10.1016/j.erss.2014.07.007
57. Torstensson, D., Wallin, F.: Potential and barriers for demand response at household customers. Energy Procedia **75**, 1189–1196 (2015). https://doi.org/10.1016/j.egypro.2015.07.570
58. Fell, M.J., Shipworth, D., Huebner, G.M., Elwell, C.A.: Exploring perceived control in domestic electricity demand-side response. Technol. Anal. Strat. Manag. **26**(10), 1118–1130 (2014). https://doi.org/10.1080/09537325.2014.974530
59. Moser, C.: The role of perceived control over appliances in the acceptance of electricity load-shifting programmes. Energy Effic. **10**(5), 1115–1127 (2017). https://doi.org/10.1007/s12053-017-9508-5
60. Stenner, K., Frederiks, E.R., Hobman, E.V., Cook, S.: Willingness to participate in direct load control: the role of consumer distrust. Appl. Energy **189**, 76–88 (2017). https://doi.org/10.1016/j.apenergy.2016.10.099
61. Yang, X., Zhang, Y.: A comprehensive review on electric vehicles integrated in virtual power plants. Sustain. Energy Technol. Assess. **48**, 101678 (2021). https://doi.org/10.1016/j.seta.2021.101678
62. Axsen, J., Goldberg, S., Bailey, J.: How might potential future plug-in electric vehicle buyers differ from current "Pioneer" owners? Transp. Res. Part D: Transp. Environ. **47**, 357–370 (2016). https://doi.org/10.1016/j.trd.2016.05.015
63. Hardman, S., Fleming, K.L., Khare, E., Ramadan, M.: A Perspective on Equity in the Transition to Electric Vehicles. MIT Science Policy Review 2021 August 20
64. Vassileva, I., Campillo, J.: Adoption barriers for electric vehicles: experiences from early adopters in Sweden. Energy **120**, 632–641 (2017). https://doi.org/10.1016/j.energy.2016.11.119
65. Westin, K., Jansson, J., Nordlund, A.: The importance of socio-demographic characteristics, geographic setting, and attitudes for adoption of electric vehicles in Sweden. Travel Behav. Soc. **13**, 118–127 (2018). https://doi.org/10.1016/j.tbs.2018.07.004
66. Lee, R., Brown, S.: Social & locational impacts on electric vehicle ownership and charging profiles. Energy Rep. **7**, 42–48 (2021). https://doi.org/10.1016/j.egyr.2021.02.057
67. Fleming, K.L.: Social equity considerations in the new age of transportation: electric, automated, and shared mobility. J. Sci. Policy Governance (2018)
68. Schmalfuß, F., Mair, C., Döbelt, S., et al.: User responses to a smart charging system in Germany: battery electric vehicle driver motivation, attitudes and acceptance. Energy Res. Soc. Sci. **9**, 60–71 (2015). https://doi.org/10.1016/j.erss.2015.08.019
69. Delmonte, E., Kinnear, N., Jenkins, B., Skippon, S.: What do consumers think of smart charging? Perceptions among actual and potential plug-in electric vehicle adopters in the United Kingdom. Energy Res. Soc. Sci. **60**, 101318 (2020). https://doi.org/10.1016/j.erss.2019.101318
70. Will, C., Schuller, A.: Understanding user acceptance factors of electric vehicle smart charging. Transp. Res. Part C: Emerging Technol. **71**, 198–214 (2016). https://doi.org/10.1016/j.trc.2016.07.006

71. Sovacool, B., Axsen, J., Kempton, W.: The future promise of Vehicle-to-Grid (V2G) integration: a sociotechnical review and research agenda. Annu. Rev. Environ. Resour. **42**(1), 377–406 (2017)
72. Bailey, J., Axsen, J.: Anticipating PEV buyers' acceptance of utility controlled charging. Transp. Res. Part A: Policy Pract. **82**, 29–46 (2015). https://doi.org/10.1016/j.tra.2015.09.004
73. Tamis, M.R., Van den Hoed, T.R.H.: Smart charging in the Netherlands. In: Proceedings of the European Battery (2017)
74. Huber, J., Schaule, E., Jung, D., Weinhardt, C.: Quo vadis smart charging? A literature review and expert survey on technical potentials and user acceptance of smart charging systems. WEVJ **10**(4), 85 (2019). https://doi.org/10.3390/wevj10040085
75. Asensio, O.I., Delmas, M.A.: The dynamics of behavior change: evidence from energy conservation. J. Econ. Behav. Organ. **126**, 196–212 (2016). https://doi.org/10.1016/j.jebo.2016.03.012
76. Al-Turjman, F., Altrjman, C., Din, S., Paul, A.: Energy monitoring in IoT-based ad hoc networks: an overview. Comput. Electr. Eng. **76**, 133–142 (2019). https://doi.org/10.1016/j.compeleceng.2019.03.013
77. Niu, S., Pan, W., Zhao, Y.: A BIM-GIS integrated web-based visualization system for low energy building design. Procedia Eng. **121**, 2184–2192 (2015). https://doi.org/10.1016/j.proeng.2015.09.091
78. Pozas, B.M., Gamero, I.A., Domínguez, A.S., de Castro, P.B.G.: A methodology to improve energy efficiency and comfort conditions with low-cost ICTs in rural public buildings. Sustain. Cities Soc. **60**, 102156 (2020). https://doi.org/10.1016/j.scs.2020.102156
79. Zou, H., Zhou, Y., Yang, J., Spanos, C.: Towards occupant activity driven smart buildings via WiFi-enabled IoT devices and deep learning. Energy Build. **177**, 12–22 (2018). https://doi.org/10.1016/j.enbuild.2018.08.010
80. Zou, H., Yuxun, Z., Yang, J., Spanos, C.: Device-free occupancy detection and crowd counting in smart buildings with WiFi-enabled IoT. Energy Build. **174**, 309–322 (2018). https://doi.org/10.1016/j.enbuild.2018.06.040
81. Khajenasiri, I., Estebsari, A., Verhelst, M., Gielen, G.: A review on internet of things solutions for intelligent energy control in buildings for smart city applications. Energy Procedia **111**, 770–779 (2017). https://doi.org/10.1016/j.egypro.2017.03.239
82. Hackbarth, A., Löbbe, S.: Attitudes, Preferences, and Intentions of German Households Concerning Participation in Peer-to-Peer Electricity Trading. Universitätsbibliothek Tübingen, Tübingen (2018)
83. Bugden, D., Stedman, R.: A synthetic view of acceptance and engagement with smart meters in the United States. Energy Res. Soc. Sci. **47**, 137–145 (2019). https://doi.org/10.1016/j.erss.2018.08.025
84. Brown, C.J., Markusson, N.: The responses of older adults to smart energy monitors. Energy Policy **130**, 218–226 (2019). https://doi.org/10.1016/j.enpol.2019.03.063
85. Balta-Ozkan, N., Davidson, R., Bicket, M., Whitmarsh, L.: Social barriers to the adoption of smart homes. Energy Policy **63**, 363–374 (2013). https://doi.org/10.1016/j.enpol.2013.08.043
86. Gumz, J., Fettermann, D.C.: What improves smart meters' implementation? A statistical meta-analysis on smart meters' acceptance. SASBE **11**(4), 1116–1136 (2022). https://doi.org/10.1108/SASBE-05-2021-0080
87. Alkawsi, G.A., Ali, N., Baashar, Y.: An empirical study of the acceptance of IoT-based smart meter in Malaysia: the effect of electricity-saving knowledge and environmental awareness. IEEE Access **8**, 42794–42804 (2020). https://doi.org/10.1109/ACCESS.2020.2977060
88. Gumz, J., Fettermann, D.C., Sant'Anna, Â.M.O., Tortorella, G.L.: Social influence as a major factor in smart meters' acceptance: findings from Brazil. Results Eng. **15**, 100510 (2022). https://doi.org/10.1016/j.rineng.2022.100510

89. Chadwick, K., Russell-Bennett, R., Biddle, N.: The role of human influences on adoption and rejection of energy technology: a systematised critical review of the literature on household energy transitions. Energy Res. Soc. Sci. **89**, 102528 (2022). https://doi.org/10.1016/j.erss.2022.102528
90. Büscher, C., Sumpf, P.: "Trust" and "confidence" as socio-technical problems in the transformation of energy systems: BioMed Central; 2015 [cited 2022 November 18] Available from: https://doi.org/10.1186/s13705-015-0063-7
91. Wilson, C., Hargreaves, T., Hauxwell-Baldwin, R.: Benefits and risks of smart home technologies. Energy Policy **103**, 72–83 (2017). https://doi.org/10.1016/j.enpol.2016.12.047

Seeing the Localization of the SDGs from a Historical Perspective: Tracing the Gradual Change of an Initiative in Japan

Jingyuan Wu(✉)

Graduate School of Frontier Sciences, The University of Tokyo, Tokyo 1138656, Japan
wujy5@g.ecc.u-tokyo.ac.jp

Abstract. It is well known that localizing the SDGs is essential and considerable research has investigated the policy translation after introducing the SDGs. However, since few have explored how the domestic policies used to be, changes brought by the SDGs are ambiguous. This article traces the gradual changing of an initiative in Japan, which evolved into the SDGs Future Cities Initiative. By analyzing official records from 2008 to 2018, this article shows that under the constant tension between international norms and local issues, the way of localizing the SDGs was shaped far before it was introduced. The institutional development of the initiative led to the SDGs becoming a political tool for enhancing existing policy and finding solutions for local issues. This article suggests that for addressing local issues, policymakers need to be flexible in using the SDGs to enhance domestic policies instead of just aiming at achieving them.

Keywords: Gradual Change · Historical Institutionalism · SDGs

1 Introduction

Most issues addressed by the United Nations 2030 Global Development Agenda are familiar, and many countries already had sustainable development-related policies even before anyone thought of the SDGs (Sustainable Development Goals). As Elder and Bartalini [1] stated, when comparing the national action plans of G20 countries, it is not easy to identify which policy is new. Consequently, the significance of whether countries created an SDGs action plan is also not fully clear. Though it is still essential to recognize that the efforts of those countries developing specific SDG plans at least indicate an intention to prioritize and publicize them, one question remains. To what extent do the SDGs make things different? This skepticism links to the critique of the SDGs. Telleria and Garcia-Arias [2] conducted a critical discourse analysis of the 2030 Global Development Agenda. They argued that it is a "fantasmatic narrative," which makes sustainable development a hollow symbol and naturalizes and consolidates the existing status quo. The scholars can only evaluate the validity of this argument after having evidence of how much domestic policies changed compared to the early ones before introducing the SDGs.

J. Baeyens et al. (Eds.): ESRE 2023, pp. 201–214, 2025.
https://doi.org/10.1007/978-981-96-4345-5_15

To contribute to scholarship in this field, this article illustrates and analyzes the gradual change of the SDGs Future Cities Initiative in Japan to provide an explanation of how and to what extent the SDGs change domestic policy. The SDGs Future Cities Initiative for regional revitalization is one of the three pillars of Japan's first national action plan for implementing the SDGs. It started in 2018 but, as will be analyzed later in this article, originated from another initiative in 2008. This paper begins by presenting the existing approaches to studying the diffusion of the SDGs and reviewing their limits. To overcome these limits, this article will illustrate the gradually changing process of the SDGs Future Cities initiative through discourse analysis and content analysis of official records. Finally, the article will conclude by contrasting my argument with existing literature about localizing SDGs.

The changing process of an initiative described here is case-specific, but this study emphasizes how pre-existing policies and different international norms shape current implementation efforts. Domestic stakeholders have leveraged the SDGs to enhance existing domestic policies to address local issues, even if those issues are not explicitly covered by the SDGs. As the analysis in this article demonstrates, the SDGs offer considerable possibilities for shaping domestic policies from local actors' perspectives, which extends beyond how faithfully a country implements the SDGs.

2 Literature Review

Scholarly interest in the localization of SDGs has risen significantly even before the release of it. Research tends to focus more on technical and managerial issues, such as developing local targets and indicators of SDGs (ex. [3, 4]), while the primary interest of this article is political. A growing body of political analysis has helped unpack the process of translating SDGs into domestic policies by using the norm diffusion theory as the conceptual framework. An official agreement assigned by the United Nations can be considered a norm, according to the definition of the 3rd type of norm by Hansen-Magnusson et al. [5]. Referring Fukuda-Parr and Hulme's [6] description of the MDGs, the SDGs could be considered a complex norm as well.

Norm diffusion refers to the international community's understanding of international norms and influences nations' domestic policies and institutions [7]. Since the 1990s, constructivists have developed two major approaches to characterizing and understanding the spread of norms: norm socialization and norm localization. Norm socialization describes a state as a 'norm taker' interested in 'belonging to the club' and adapting their behavior to that club's rules [8]. A classic theory examining how norms affect political change is the norm's "life cycle" [9]. In this theory, according to a "logic of appropriateness," states "become socialized to accept…values, rules, and roles. They internalize the roles and rules as scripts to which they conform, not out of conscious choice, but because they understand these behaviors to be appropriate" [10]. When talking about why the Japanese government has taken action to implement the SDGs, Japanese scholar Kanie [11] claims that it is the universal goal endorsed by the member states of the United Nations. This kind of claim emphasizes the norm-promotion activities of international actors and the influence of the SDGs on state identities, while downplaying the domestic interests and political process of norm adoption.

The second approach, norm localization, which emphasizes domestic factors, provides only superficial process tracing. According to Acharya's [12] and Wiener's [13] studies, domestic actors strategically reinterpret and reshape international norms to a new context to build congruence with existing domestic policy in the process of localizing norms. However, one of the main objectives of Japan's SDG Future Cities Initiative is to deal with depopulation and revitalize local municipalities' economies, which are not included in the SDGs. It seems that congruence between the SDGs and existing domestic policy is unnecessary.

To overcome the limits of the former two approaches, Zimmermann [14] introduced a more detailed image of the "cycles of translation." In every new context, norms and institutional models are translated in more local and international contexts in the sequence shape and constraint leeway for translation in future stages. A study on the SDGs by Hustad [15] is derived from this theory. By focusing on the policy translation process, Hustad highlighted the discursive struggles in the translation processes of SDG17 and explained why a particular institutional model is taken up in discourse and is interpreted [15]. However, since the study scope only contains domestic contestation after the SDGs were released, to what extent it differs from the former policy is still a blur.

In brief, most studies started the policy translation narrative from the process of making a national plan for implementing the SDGs and pointing to the dynamic nature of the SDGs as open to change due to translation processes in new contexts. However, little attention has been given to the pre-SDGs phase, leaving the extent of domestic policy changes resulting from introducing the SDGs into the domestic context ambiguous. To gain a clearer understanding of the complex impact of the SDGs and reflect on skepticism about their efficacy, it is necessary to include the pre-SDGs phase in the policy analysis. Thus, this article extends the analysis timeline to trace tensions in the domestic political process under multiple norms and how they eventually shaped the SDGs Future Cities Initiative.

3 Method and Data

This study was designed as an exploratory single-case study to investigate how an initiative has evolved in response to changing international trends while taking root in Japan's social context. The analysis examines the period from 2008 to 2018 and divides the initiative's evolution into three distinct stages: the Environmental Model Cities Initiative (commencing in 2008), the Eco-Future Cities Initiative (commencing in 2011), and the SDGs Future Cities Initiative (commencing in 2018). The presentation of the three stages in Fig. 1 is intentional, as the three stages were not linear switching but are more appropriately understood as an extension process of its scope.

To reveal the dynamic changing, this article conducted process tracing, which analyzes sequences of events to explore the relevant linking mechanisms and intervening processes in specific cases. The approach of this article combines discourse analysis and content analysis based on data collection in Japanese. I applied discourse analysis to the official meeting, including consultation and speeches about the initiative. The aim is to uncover the interactions between international and domestic factors and the tension between different actors during policymaking.

Fig. 1. The 3 stages of the initiative's evolution. *Source* Made by author

The official meeting documents and speeches included in this analysis were carefully selected. As a national policy, the Diet repeatedly mentions the initiative. Searching "Environmental Model Cities," "Eco-Future Cities," and "SDGs Future Cities" in Japanese on the Diet online documenting system (https://kokkai.ndl.go.jp/#/), there are a total of 112 meeting records hit by the date of August 13, 2022. Also, I gathered official speeches for the annual symposium from 2008 to 2018, featured on the Environment Future Cities initiative Portal, and official meeting records for the SDGs Future Cities from 2015 to 2016, featured on the Regional Revitalization Portal Site (https://www.chisou.go.jp/sousei/index.html).

Then, to reveal the similarities and differences in thematic occurrences across the 3 evolution stages of the initiative, my content analysis scrutinizes official application guidelines and official reports of selected cities issued by the Japanese Cabinet Office. This article focused on three perspectives: definitions, objects of the initiative, and application requirements. By tracing how the 3 perspectives changed, this article will reveal the progressive evolution of the initiative over time and to what extent the SDGs changed it.

4 The Evolution of SDGs Future Cities Initiative

4.1 Stage 1: Environmental Model Cities Initiative

A new initiative started under the background that Japan was required to reduce its greenhouse gas emissions by 6% between 2008 and 2012 under the Kyoto Protocol. Shortly after the Thirteenth session of the Conference of the Parties to the UNFCCC (United Nations Framework Convention on Climate Change), 18 January 2008, during the 169th Diet, Prime Minister of Japan Yasuo Fukuda delivered a policy speech on building a low-carbon society and officially announced the Environmental Model Cities Initiative for the first time.

In cooperation with local municipalities, 10 cities will be selected as Environmental Model Cities, which will set high targets and take on the challenge of pioneering initiatives, such as significant reductions in greenhouse gas emissions [16].

It is noteworthy that efforts towards building a low-carbon society, including the Environmental Model Cities Initiative, are intensely conscious of Japan's leadership role in the international community. In the speech, Prime Minister Fukuda showed his ambition of "leading the international community by promoting the transition to a low-carbon society that sets a precedent for the world" [16]. Fukuda stressed that the world is facing a climate change crisis and that Japan's technologies to save energy and protect the environment are powerful tools in the fight against this crisis. He also pointed out that the G8 Summit in July 2008 was an excellent opportunity to show Japan's commitment to solving environmental issues [16]. At the same Diet sessions, the Minister [17] and the Vice-Minister of Environment [18] gave a consistent account of the initiative in line with Prime Minister Fukuda's proposal.

The emphasis on emission reduction is reflected in the requirements for the applications for the initiative issued by the Cabinet Office in 2008, in which the very first of the criteria for Environmental Model Cities is to "set a goal of significantly reducing greenhouse gas emissions" [19]. Specifically, the official call for proposals recommends the following:

Aim at a long-term target of more than halving emissions by 2050.

Aim to peak out emissions within cities and regions as soon as possible.
Aim for at least 30% energy efficiency improvements by 2020 [19].

Although the initiative places such emphasis on aggressive climate change efforts, the operation of the initiative is carried out by the Regional Revitalization Headquarters under the Cabinet Office, instead of the environmental department. More specifically, model cities were selected by the Environmental Model Cities and Low-carbon Society Subcommittee, which was set up under the expert panel on global warming in March 2008. While statements by Prime Minister Fukuda and the Ministry of Environment set the tone for the initiative, which is to pursue emissions reductions as a responsible member of the global community, the actors in the Cabinet Office have the authority to run the initiative and show different views on the purpose of the initiative.

The initiative was positioned as a "new direction of action for urban issues" in the Plan for the Development of Cities and Livelihoods, decided by the Regional Revitalization Headquarters on 29 January 2008. The plan's description of the initiative, although highlighting low carbon emissions reduction, is silent on fulfilling international commitments but focuses on achieving sustainable and comfortable urban life [20]. The minister in charge of decentralization reform and regional regeneration explained the initiative's purpose at the Diet on 22 February 2008, as "taking pioneering actions in new lifestyles, urban and transport systems" [21].

The purpose of promoting a new way of life became even more vital during the discussion in the Environmental Model Cities and Low-carbon Society Subcommittee. During the Subcommittee meeting, there are three strategies for building the legitimacy of the purpose of valuing citizens' life quality. The first is claiming that the purpose is emotionally proper or desirable. Commissioner Ishida said, "We definitely want to create a low carbon while highly satisfying society" [22]. Multiple commissioners thought the word "low carbon" sounded too negative, and the initiative might send a negative message that people have to endure it [23]. As Commissioner Masumoto said, "It is not good to create spaces where people are not comfortable" [22].

The second strategy is rationalizing the purpose by mentioning past failures to pursue emission reduction without properly considering citizens' demands. Commissioner Murakami pointed out that energy consumption in the civilian sector is directly related to citizens' lives and that although various measures have been taken, they have not successfully curbed it sufficiently. Therefore, he urged that lifestyle innovation is needed to pursue the initiative [22].

The third one is legitimizing the purpose by claiming it as a necessary factor in improving the initiative's effects. There was a view that the initiative must emphasize the compatibility with economic development to "sounds like a positive message" to the citizens and "have a reach that will change people's behavior" [22]. Especially for those economically depressed and depopulated areas, citizens' willingness to participate must be encouraged by creating a positive image of the initiative [23]. Therefore, the Subcommittee suggested that the call for application should include expressions such as the initiative are for "the whole region will become economically active" and "the next generation will actually become affluent" [23].

The meaning of model "city" mentioned by Prime Minister Fukuda, as well as the emphasis on solving urban issues by the Department of Environment and the Cabinet Office, was also partly revised by the Regional Revitalization Headquarters and the Subcommittee. When the staff from the Regional Revitalization Headquarters presented the draft call to the Subcommittee, he said, "In making the selection… we will carefully consider the balance between regions, the balance between city sizes, and the actual emissions in different cities" [22]. Some commissioners expressed concern about further expanding the range of candidates. Commissioner Murakami said, "I think we should assume that the applicants range from large ordinance-designated cities to very small municipalities" [22]. Also, Commissioner Masumoto said, "Municipalities applying for the Initiative will not only just those have large energy consumption," and looked for that the selection criteria of the Environment Cities will give consideration to diverse characteristics of local municipalities [22].

In fact, a total of 23 municipalities were selected as Environmental Model Cities in three rounds of competition in 2008, 2012, and 2013, of which 6 municipalities had populations less than 50,000 [24]. That said, during the actual selection procedure, the candidates were extended from "Cities" to small-scale local municipalities, even in rural areas.

To sum up, during stage 1, despite the preeminence of the notion of reducing emissions in urban areas at the onset of the Kyoto Protocol, the Regional Revitalization Headquarters and the Subcommittee actively redefined the initiative's content. The initiative's inclusivity and moderation increased, allowing more small local municipalities to participate.

4.2 Stage 2: Eco-Future Cities Initiative

During the first binding period of the Kyoto Protocol, although Japan was under pressure to meet the target for significant emission reductions, the Lehman crisis in 2009 made the tension between economic growth and countermeasures against the greenhouse effect more intense. To stimulate the domestic economy, in December 2009, the Cabinet of Yukio Hatoyama decided on a basic policy outline called the New Growth Strategy. Later, it was released as a national strategy in 2010 by the new Cabinet led by Naoto Kan. The national strategy named a new Initiative, the Eco-Future Cities Initiative, as one of the national strategic projects [25].

The New Growth Strategy stated the purpose of the Eco-Future Cities Initiative as "creating and promoting the world's top success stories of future-oriented technologies, systems, services, and urban planning" [25]. From the statement, we could see the purpose of drastic reduction of emissions, clearly emphasized in the Environmental Model Cities initiative, faded from the initial concept of the Eco-Future city. There is a certain ambiguity in the positioning of the Eco-Future Cities Initiative, which triggered a divergence of opinions in a high-level meeting.

The Cabinet Office arranged a new group of experts named the Eco-Future Cities Initiative Experts Panel. It held five meetings from October 2010 to February 2011. While the issue of greenhouse gas emissions remained one of the priorities at the conference, members expressed concern about Japan's declining population and aging amid an economic downturn, such as "People won't have happiness just by eliminating CO_2 emissions." It is said that building a low-carbon society should not be the goal of the initiative but rather "a means" to "improve the sustainability and creativity of society" [26]. And it is impossible to solve all urban problems in this initiative, which should focus on "new growth" "internationalization" and "super-aging" [27]. Due to the differences mentioned above in purpose and areas of concern, there was a suggestion to separate the new Eco-Future Cities Initiative from the old one and to operate these two in "a completely different scheme" [27].

After the expert meetings, the Cabinet Office wrote an executive summary of the discussion on the new initiative, and took it to the Evaluation and Advisory Group of the Environmental Model Cities Initiative, which contains most of the former members from the Subcommittee in Stage 1. The Cabinet Office introduced the Eco-Future Cities Initiative as a different initiative, no longer only focusing on building a low carbon society [28]. The Evaluation and Advisory Group members showed their concerns. Commissioner Yakushiji said, "If you put too many elements into the future city, it will make sense of connectedness hollow. Setting a clear purpose may narrow the scope, but it will likely be implemented. By adding various elements, the scope will be wider, but the possibility of implementation will be low" [28]. Commissioner Murakami, the leader of the Evaluation Proposal Group and the former Subcommittee, agreed with Yakushiji's comment. In line with such concerns, there are plural views from the Evaluation and Advisory Group that the government should operate the two initiatives separately [28].

When Murakami, the leader of the advisory group, first publicly introduced the Eco-Future Cities Initiative at a promotion forum held in April 2011, there was no information about the Environmental Model Cities Initiative in his presentation material [29]. The same is true for the call for applications at the time it was first released [30]. All above reflects the initial separation of the Eco-Future City Initiative and the Environmental Model City Initiative.

In the application guidelines, the Eco-Future City was defined as "a town that creates environmental, social and economic values to achieve," "a town where everyone wants to live," and "a town where everyone has vitality" [30]. The initiative no longer aims to lower carbon emissions, and there is no requirement for submitting a significant emissions reduction plan. The initiative aims to "create a new city with new values in the center of the world for the environment and super-aging." Under this goal, the applicants "must include proposals related to the environment and super-aging" in their plans [30].

However, the Great East Japan Earthquake on March 11, 2011, significantly impacted Japan's energy supply and greenhouse effect countermeasures. When Japan used all possible resources to support the disaster area, the Eco-Future Cities Initiative selection procedure changed. The initiative added set-aside quotas for the Great East Japan Earthquake disaster areas. 6 of the 11 cities selected in 2011 were disaster areas. The others are 4 Environmental Model Cities selected in 2008, and Kashiwa City is the only new entry [24].

Consequently, in May 2012, the Promotion Council for Low Carbon City, led by the Subcommittee of the former initiative, was reorganized into the Promotion Council for Eco-Future Cities Initiative. The statement of the reorganization illustrated the relationship between the two initiatives.

> Since the "Eco-Future City" is an essential response to the environment and super-aging society, it can be positioned as a further development of the Environmental Model Cities Initiative that has led the way in creating an environmental and low-carbon city. The past efforts can be effectively utilized to promote the Eco-Future Cities Initiative [31].

The two initiatives merged eventually. The new one was re-positioned as the upgrade version of the old one, with more consideration about social issues such as super-aging. Since the initiative aimed at creating and promoting best practices to the world, it was listed as one of the Japanese government's inputs to Rio +20 [32]. The Ministry of Foreign Affairs also held a themed event during the Rio +20 [33]. But it doesn't mean the government pursues the initiative more actively domestically. The initiative no longer called for new applications after 2013.

During Stage 2, the newly introduced Eco-future City Initiative initially displayed tendencies of diverging from the previous Environmental Model Cities Initiative. However, because of a change in government and the impact of the East Japan Earthquake, the institutional framework of the initiative underwent a significant transformation. The Eco-future Cities Initiative was subsequently redefined as an upgraded version of the former one, with its scope broadened from environmental issues to encompass more expansive sustainable development concerns. The expansion of the definition opened the way for it to link to the SDGs in the next stage.

4.3 Stage 3: SDGs Future Cities Initiative

When the SDGs were launched in 2015, the Promotion Council for Eco-Future Cities Initiative already had members from over a hundred local municipalities [34]. The Promotion Council even took the lead in organizing a working group on the guidelines for local municipalities to implement the SDGs [35]. Though this initiative was supposed to end in March 2017.

Looking for a new start, the keyman Murakami, who had led the Subcommittee in Stage 1 and the Evaluation and Advisory Group in Stage 2, proposed the idea of merging the initiative with the SDGs. He brought the proposal of merging the SDGs into the policy for regional revitalization as a new initiative into the SDGs Roundtable hosted by the Cabinet Office in May 2017. In Murakami's proposal, the new initiative aims to revitalize local societies from the three aspects of the environment, social welfare, and economy [36]. The Regional Revitalization Headquarters showed their support and said in the SDGs Promotion Headquarters meeting, "We will further develop the Eco-Future Cities Initiative and formulate measures to achieve the SDGs in the local municipalities. By actively promoting these measures, we will achieve further development of local societies" [37].

So-called regional revitalization has a unique character deeply entrenched in Japanese society. Japan's population has been continuously declining. After 2011, Public concern was highly aroused by some popular discourse such as "the disappearance of local municipalities" [38]. Under such pressure, the Japanese government has launched a package plan named "regional revitalization" to halt population decline in local municipalities and the concentration of population in Tokyo. As the start, in 2014, the Japanese government released a policy outline and a comprehensive national strategy, which began to be fully operational in 2016. Though the SDGs contain no content precisely about dealing with population decline, merging the SDGs and the regional revitalization policy seems natural. After all, the old one had already covered local sustainability issues.

The proposal of the SDGs Future Cities Initiative was also in line with the Ministry of Foreign Affairs plan. The Ministry of Foreign Affairs was preparing to host the G20 Summit, the Tokyo International Conference on African Development, and the High-Level Political Forum in 2019. They planned to show Japan's characteristics to the international society by presenting a Japan's SDGs model. Because of it, as the Ministry of Foreign Affairs explained at the 4th meeting of the SDGs Roundtable, the government planned to make a new national strategy for the SDGs after the first SDGs implementation plan in 2016 [35]. Eventually, The Ministry of Foreign Affairs decided the 3 main themes for the Japanese SDGs model presented to the world: regional revitalization, society 5.0, and empowerment of women and youth [39]. And it was written into the SDGs Action Plan 2018, released in December 2017 [40]. According to the definition by the Cabinet Office, the SDGs Future Cities should be as follows.

> Cities and towns selected promote primary and comprehensive initiatives in line with the principles of the SDGs, especially those with high potential to realize sustainable development through creating new value in the three aspects of the economy, society, and environment [41].

The expression of the initiative is highly like the definition of Eco-Future Cities, which also contains the expression as "meeting a certain level concerning the environment, society, and economy, and creating further value in the 3 aspects" [30]. Nevertheless, the requirements for the application are softer than the old initiatives, as shown in Table 1. Since the first initiative, Environmental Model Cities Initiative was a project created with the implementation of greenhouse gas reduction targets in mind, it required local municipalities to set significant reduction targets. The second one, Eco-Future Cities Initiative, didn't require setting strict reduction targets, but it did narrow the scope down to actions related to the environment and super-aging issues. In contrast, although the new SDGs Future Cities Initiative requires the selected local municipalities to establish indicators and monitor the progress, it allows the local municipalities to design their own local indicators [41]. According to the application guidelines, as long as the local municipalities are meant to take action to make the local economy, environment, and society better, which is a very large and vague requirement, they could apply for this initiative [41].

The lower difficulty of application reflects in the higher selection ratio under the SDGs Future Cities Initiative, which can be observed in Table 2. The Environmental Model Cities Initiative selected 16 municipalities out of 115 [42], and the Eco-future Cities Initiative chose only 11 municipalities out of 92 [43]. In contrast, the SDGs Future Cities Initiative has selected about 30 local municipalities annually from 2018 to 2022 [44]. Once chosen as the SDGs Future City, the local municipalities not only get financial support but also could utilize the potential of the SDGs brand. According to a national survey conducted by a committee under the Cabinet Office, when local municipalities were asked why they applied for the initiative, the top reason in 2021 was "to strengthen brand power," in a ratio of 40% [45].

Table 1. Main contents of each initiative

	Environmental Model Cities	Eco-Future Cities	SDGs Future Cities
Concept	Low-Carbon	focusing on the environmental issues and super-aging problem	Sustainable development in three dimensions: economic, social, and environmental
Requirement for applicants	Large reduction in greenhouse gas emission	Actions on solving environmental issues and super-aging problem	Integrated actions on economy, social welfare, and environment

Source Made by author based on the application guidelines for each initiative [19, 30, 41]

As mentioned above, far before the SDGs launched, the old initiatives evolved with more enriched content related to sustainable development, and the institution expanded the network to local municipalities. All the institutional development ultimately led to the emergence of the SDGs Future Cities Initiative, which integrated regional revitalization

Table 2. Selection ratios of each initiative by the fiscal year 2022

	Environmenta Model Citiesl	Eco-Future Cities	SDGs Future Cities
No. of applications	115	92	279
No. of selections	16	11	154
Chance of being selected	14%	12%	55%

Source Made by author based on the data from the Regional Revitalization Portal Site [44]

into the SDGs action plan. Without the prior development of these initiatives, it is difficult to imagine that policies addressing local depopulation and economic stagnation, which are not precisely mentioned by the SDGs, would have become a crucial pillar of the SDGs action plan. Furthermore, the integration of the international goal also transformed the old initiative into a more diverse and adaptable one, providing it with greater legitimacy and recognition.

5 Conclusion

This article revealed that the SDGs Future Cities Initiative is more than a result of socializing nor localizing the SDGs but a stage upon the much longer progress of the gradual institutional change.

The first environmental Moder Cities Initiative was launched as a policy centered on reducing carbon emissions when the first binding period of the Kyoto Protocol approached. But the Regional Revitalization Headquarters and the experts' group, which operates the initiative, emphasize the duality of emissions reductions and people's living. They reshaped the initiative's purpose more openly to local municipalities that are not cities with high carbon emissions. With changes in the government, the impact of a major earthquake, and the emergence of the SDGs, the initiative evolved further, expanding its scope to encompass broader sustainable development issues. Eventually, the experts who had been operating the former initiatives for years utilized the SDGs' influence and legitimacy to extend the initiative's benefits to local towns facing sustainable problems such as depopulation. The SDGs is just a single step in a much longer institutional development process.

The SDGs Future Cities Initiative in Japan is not only an attempt to implement the SDGs but also a way to address local issues by leveraging the SDGs' popularity. It would be difficult to reveal the significance of this policy if someone were to measure it simply by the content and indicators of the original SDGs, because it is rooted in the institutional development and the context of unique social issues in Japan. This suggests that policymakers need to be flexible in using international norms to enhance domestic policies. It also implies that new policies aimed only at achieving the SDGs may not be sufficient to address local issues effectively.

It is worth mentioning that the approach that favors strengthening existing domestic policies also has drawbacks. It could lead to neglecting long-standing but underappreciated problems that are not part of the main policy agenda. This problem might also occur in Japan but is out of this article's scope. Future research agendas shall shed further light on how to take the balance between international commitments and local issues instead of creating more and more technical tools for measuring global goal progress.

References

1. Elder, M., Bartalini, A.: Assessment of the G20 countries' concrete SDGs Implementation Efforts: Policies and Budgets Reported in their 2016–2018 Voluntary National Reviews. Institute for Global Environmental Strategies (2019)
2. Telleria, J., Garcia-Arias, J.: The fantasmatic narrative of 'sustainable development': a political analysis of the 2030 global development agenda. Environ. Plann. C: Polit. Space **40**(1), 241–259 (2021)
3. Reddy, P.: Localising the sustainable development goals: the role of local government in context. Afr. J. Public Aff. **9**(2), 1–15 (2016)
4. Valencia, S.C., et al.: Adapting the sustainable development goals and the new urban agenda to the city level: initial reflections from a comparative research project. Int. J. Urban Sustain. Dev. **11**(1), 4–23 (2019)
5. Hansen-Magnusson, H., Vetterlein, A., Wiener, A.: The problem of non-compliance: knowledge gaps and moments of contestation in global governance. J. Int. Relat. Dev. **23**(3), 636–656 (2020)
6. Fukuda-Parr, S., Hulme, D.: International norm dynamics and the end of poverty: understanding the millennium development goals. Glob. Gov. **17**, 17–36 (2011)
7. Checkel, J.T.: International norms and domestic politics: bridging the rationalist-constructivist divide. Eur. J. Int. Rel. **3**(4), 473–495 (1997)
8. Flockhart, T.: Complex socialization and the transfer of democratic norms. In: Flockhart, T. (ed.) Socializing Democratic Norms: The Role of International Organizations for the Construction of Europe, pp. 43–62. Springer, Heidelberg (2005)
9. Finnemore, M., Sikkink, K.: International norm dynamics and political change. Int. Organ. **52**(4), 887–917 (1998)
10. Finnemore, M.: Norms, culture, and world politics: insights from sociology's institutionalism. Int. Organ. **50**(2), 325–347 (1996)
11. Kanie, N.: SDGs. Chuokoron-Shinsha Press, Tokyo (2020) (Japanese)
12. Acharya, A.: How ideas spread: Whose norms matter? Norm localization and institutional change in Asian regionalism. Int. Organ. **58**(2), 239–275 (2004)
13. Wiener, A.: The dual quality of norms and governance beyond the state: sociological and normative approaches to 'interaction.' Crit. Rev. Int. Soc. Pol. Phil. **10**(1), 47–69 (2007)
14. Zimmermann, L.: Global Norms with A Local Face: Rule-of-Law Promotion and Norm Translation. Cambridge University Press (2017)
15. Hustad, O.: From global goal to local development policy: how partnerships as a policy idea changes through policy translation. Dev. Policy Rev. **41**(2), e12659 (2023)
16. Fukuda, Y.: Speech at the 169th Diet, House of Representatives, Plenary Session No. 1 meeting, January 18, 2008, https://kokkai.ndl.go.jp/txt/116905254X00120080118/9, last accessed 21 Jan. 2023 (Japanese)
17. Kamoshita, I.: Speech in the 169th Diet, House of Representatives, Environment Committee, No. 1 meeting, February 22, 2008, https://kokkai.ndl.go.jp/txt/116904006X00120080222/4, last accessed 21 Jan. 2023 (Japanese)
18. Sakurai, I.: Speech at the 169th Diet, House of Representatives, Environment Committee, No. 2, March 18, 2008, https://kokkai.ndl.go.jp/txt/116904006X00220080318/7, last accessed 21 Jan. 2023 (Japanese)

19. Regional Revitalization Portal Site.: Environmental model city application guidelines. https://www.chisou.go.jp/tiiki/kankyo/upload/120907koubo/boshuyouryou.pdf, last accessed 21 Jan. 2023 (Japanese)
20. Regional Revitalization Headquarters.: Plan for the Development of Cities and Livelihoods. https://www.chisou.go.jp/tiiki/081219hattenplan.pdf, last accessed 21 Jan. 2023. (Japanese)
21. Yasuda, H.: Speech at the 169th Diet, House of Representatives, Cabinet Committee No. 1 meeting, February 22, 2008, https://kokkai.ndl.go.jp/txt/116904889X00120080222/8, last accessed 21 Jan. 2023 (Japanese)
22. Regional Revitalization Portal Site.: 1st Environmental Model Cities/Low-carbon Society Subcommittee Meeting on March 31, 2008, https://www.chisou.go.jp/tiiki/kankyo/upload/080331bunkakai/01gijiyousi.pdf, last accessed 21 Jan. 2023 (Japanese)
23. Regional Revitalization Portal Site.: 2nd Environmental Model Cities/Low-carbon Society Subcommittee Meeting on April 11, 2008, https://www.chisou.go.jp/tiiki/kankyo/upload/080411bunkakai/02gijiyousi.pdf, last accessed 21 Jan. 2023 (Japanese)
24. Regional Revitalization Portal Site.: Environmental Model Cities and Eco-Future Cities, https://www.chisou.go.jp/tiiki/kankyo/pdf/kankyo_gaiyo.pdf, last accessed 21 Jan. 2023 (Japanese)
25. Cabinet Office, Government of Japan.: The New Growth Strategy, https://www.kantei.go.jp/jp/sinseichousenryaku/sinseichou01.pdf, last accessed 21 Jan. 2023 (Japanese)
26. Regional Revitalization Portal Site.: The 1st Meeting of the Experts Panel for the Eco-Future Cities initiative on October 27, 2010, https://www.chisou.go.jp/tiiki/kentoukai/dai1/siryou6.pdf, last accessed 21 Jan. 2023 (Japanese)
27. Regional Revitalization Portal Site.: The 2nd meeting of the Experts Panel for the Eco-Future Cities initiative on November 26, 2010, https://www.chisou.go.jp/tiiki/kentoukai/dai3/siryou5.pdf, last accessed 21 Jan. 2023 (Japanese)
28. Regional Revitalization Portal Site.: The 2nd Environmental Model Cities Evaluation and Advisory Group Meeting on February 17, 2011, https://www.chisou.go.jp/tiiki/kankyo/upload/110217AG/gijiroku.pdf, last accessed 21 Jan. 2023 (Japanese)
29. Murakami, S.: Speech at the eco-future cities initiative promotion forum, April 24, 2011. https://doc.future-city.go.jp/doc/pdf/forum_shiryou/murakami_tokyo.pdf, last accessed 21 Jan. 2023 (Japanese)
30. Regional Revitalization Portal Site.: Eco-Future Cities Application Guideline. https://www.chisou.go.jp/tiiki/kankyo/teian/sentei2.html, last accessed 21 Jan. 2023 (Japanese)
31. Regional Revitalization Portal Site.: Further promotion of the Eco-Future Cities initiative and the expansion of the Environmental Model Cities initiatives. https://www.chisou.go.jp/tiiki/kankyo/kaigi/kako/pdf/boshu/20120522_kentoukai_ref04.pdf, last accessed 21 Jan. 2023 (Japanese)
32. Ministry of Foreign Affairs of Japan.: The Green Future initiative, https://www.mofa.go.jp/mofaj/gaiko/kankyo/rio_p20/pdfs/midori.pdf, last accessed 21 Jan. 2023 (Japanese)
33. Ministry of Foreign Affairs of Japan: Official Side Event Hosted by the Government of Japan in Rio+20 (Summary), https://www.mofa.go.jp/mofaj/gaiko/kankyo/rio_p20/side_event.html, last accessed 21 Jan. 2023 (Japanese)
34. The Promotion Council for Eco-Future Cities Initiative.: Members of the Promotion Council, https://future-city.go.jp/pclcc/about/pdf/pclcc_member300228.pdf, last accessed 21 Jan. 2023 (Japanese)
35. The Promotion Council for Eco-Future Cities Initiative.: Working Group on the City-based SDGs, https://future-city.go.jp/pclcc/WG_12/, last accessed 21 Jan. 2023 (Japanese)
36. SDGs Promotion Headquarters.: SDGs Promotion Round Table's 4th meeting on December 6, 2017, https://www.kantei.go.jp/jp/singi/sdgs/entakukaigi_dai4/gijiyousi.pdf, last accessed 21 Jan. 2023 (Japanese)

37. SDGs Promotion Headquarters, SDGs Promotion Headquarters' 3rd Meeting on June 9, 2017. https://www.kantei.go.jp/jp/singi/sdgs/dai3/gijiroku.pdf, last accessed 21 Jan. 2023 (Japanese)
38. Tanaka, T.: Sociology of Relational Population: Regional Revitalization in the Age of Declining Population. Osaka University Press, Osaka (2021) (Japanese)
39. SDGs Promotion Headquarters.: SDGs Promotion Round Table's 5th meeting on May 30, 2018, https://www.kantei.go.jp/jp/singi/sdgs/entakukaigi_dai5/gijiyousi.pdf, last accessed 21 Jan. 2023 (Japanese)
40. SDGs Promotion Headquarters.: SDGs Action Plan 2018. https://www.kantei.go.jp/jp/singi/sdgs/pdf/actionplan2018.pdf, last accessed 21 Jan. 2023 (Japanese)
41. Regional Revitalization Portal Site: SDGs Future Cities Application Guideline, https://www.chisou.go.jp/tiiki/kankyo/teian/sdgs_2022sentei.html, last accessed 21 Jan. 2023 (Japanese)
42. Regional Revitalization Portal Site.: The Selection of Environmental Model Cities Initiative, https://www.chisou.go.jp/tiiki/kankyo/teian/sentei.html, last accessed 21 Jan. 2023 (Japanese)
43. Regional Revitalization Portal Site.: The Selection of Eco-Future Cities Initiative, https://www.chisou.go.jp/tiiki/kankyo/teian/sentei2.html, last accessed 21 Jan. 2023 (Japanese)
44. Regional Revitalization Portal Site.: Regional Revitalization, and SDGs, https://www.chisou.go.jp/tiiki/kankyo/index.html, last accessed 21 Jan. 2023 (Japanese)
45. Regional Revitalization Portal Site.: The Results of National Questionnaire Survey 2021, https://www.chisou.go.jp/tiiki/kankyo/kaigi/SDGs_enquete_chousa.html, last accessed 21 Jan. 2023 (Japanese)

Author Index

J. Baeyens et al. (Eds.): ESRE 2023, pp. 215–216, 2025.
https://doi.org/10.1007/978-981-96-4345-5

The manufacturer's authorised representative in the EU is Springer Nature Customer Service Centre GmbH, Europaplatz 3, 69115 Heidelberg, Germany. If you have any concerns regarding our products, please contact ProductSafety@springernature.com

Printed and bound by CPI Group (UK) Ltd, Croydon, CR0 4YY
30/09/2026
02225297-0004